Monographs in Theoretical Computer Science
An EATCS Series

Springer
Berlin
Heidelberg
New York
Barcelona
Budapest
Hong Kong
London
Milan
Paris
Santa Clara
Singapore
Tokyo

Kurt Jensen

Coloured Petri Nets

Basic Concepts, Analysis Methods and Practical Use
Volume 1

Second Edition

With 84 Figures

Springer

Author

Prof. Kurt Jensen
Aarhus University
Computer Science Department
Ny Munkegade, Bldg. 540
DK-8000 Aarhus C, Denmark

Series Editors

Prof. Dr. Wilfried Brauer
Institut für Informatik, Technische Universität München
Arcisstrasse 21, D-80333 München, Germany

Prof. Dr. Grzegorz Rozenberg
Department of Computer Science
University of Leiden, Niels Bohrweg 1, P.O. Box 9512
2300 RA Leiden, The Netherlands

Prof. Dr. Arto Salomaa
Data City
Turku Centre for Computer Science
FIN-20520 Turku, Finland

Cataloging-in-Publication Data applied for
Die Deutsche Bibliothek - CIP-Einheitsaufnahme

Jensen, Kurt:
Coloured petri nets: basic concepts, analysis methods and practical use / Kurt Jensen. - Berlin; Heidelberg; New York; Barcelona; Budapest; Hong Kong; London; Milan; Paris; Santa Clara; Singapore; Tokyo: Springer
(Monographs in theoretical computer science)
Vol. 1.. - 2. ed., 2., corr. printing. - 1997

Second corrected printing 1997

ISBN 978-3-642-08243-6

Cover Design: MetaDesign, Berlin

Note to the Corrected Reprint

This corrected reprint is identical to the first and second edition – except for the correction of a few errors. Most of these are simple spelling mistakes, duplicated words, and other small errors that have little or no influence on the readability of the text.

Despite all efforts some errors remain. That seems to be inevitable, no matter how many people read the manuscript. If you wish to report errors or discuss other matters you may contact me via electronic mail: kjensen@daimi.aau.dk. You may also take a look at my WWW pages: http://www.daimi.aau.dk/~kjensen/. They contain a lot of material about CP-nets and the CPN tools, including a list of errata for this book.

Aarhus, Denmark
February 1997

Kurt Jensen

Preface

This book presents a coherent description of the theoretical and practical aspects of Coloured Petri Nets (CP-nets or CPN). It shows how CP-nets have been developed – from being a promising theoretical model to being a full-fledged language for the design, specification, simulation, validation and implementation of large software systems (and other systems in which human beings and/or computers communicate by means of some more or less formal rules). The book contains the formal definition of CP-nets and the mathematical theory behind their analysis methods. However, it has been the intention to write the book in such a way that it also becomes attractive to readers who are more interested in applications than the underlying mathematics. This means that a large part of the book is written in a style which is closer to an engineering textbook (or a users' manual) than it is to a typical textbook in theoretical computer science. The book consists of three separate volumes.

The first volume defines the net model (i.e., hierarchical CP-nets) and the basic concepts (e.g., the different behavioural properties such as deadlocks, fairness and home markings). It gives a detailed presentation of many small examples and a brief overview of some industrial applications. It introduces the formal analysis methods. Finally, it contains a description of a set of CPN tools which support the practical use of CP-nets. Most of the material in this volume is application oriented. The purpose of the volume is to teach the reader how to construct CPN models and how to analyse these by means of simulation.

The second volume contains a detailed presentation of the theory behind the formal analysis methods – in particular occurrence graphs with equivalence classes and place/transition invariants. It also describes how these analysis methods are supported by computer tools. Parts of this volume are rather theoretical while other parts are application oriented. The purpose of the volume is to teach the reader how to use the formal analysis methods. This will not necessarily require a deep understanding of the underlying mathematical theory (although such knowledge will of course be a help).

The third volume contains a detailed description of a selection of industrial applications. The purpose is to document the most important ideas and experiences from the projects – in a way which is useful for readers who do not yet have personal experience with the construction and analysis of large CPN diagrams. Another purpose is to demonstrate the feasibility of using CP-nets and the CPN tools for such projects.

Together the three volumes present the theory behind CP-nets, the supporting CPN tools and some of the practical experiences with CP-nets and the tools. In our opinion it is extremely important that these three research areas have been developed simultaneously. The three areas influence each other and none of them could be adequately developed without the other two. As an example, we think it would have been totally impossible to develop the hierarchy concepts of CP-nets without simultaneously having a solid background in the theory of CP-nets, a good idea for a tool to support the hierarchy concepts, and a thorough knowledge of the typical application areas.

How to read Volume 1

We assume that the reader is familiar with basic mathematical notation, such as sets and functions. We also assume some basic knowledge about computer systems and the development of such systems. No prior knowledge of Petri nets is assumed.

Readers who are interested in both the theory and practical use of CP-nets are advised to read all chapters (in the order in which they appear). Readers who primarily are interested in the practical use of CP-nets may find it sufficient to read Chap. 1, Sects. 3.1–3.2, and Chaps. 6–7. They may, however, also glance through Chap. 5 which gives a very informal introduction to the basic analysis methods. Readers who primarily are interested in the theoretical aspects of CP-nets may decide to skip Chaps. 6–7 (although this is not recommended).

For all readers it is advised to do the exercises of those chapters which are read (or at least some of them). It is also recommended to have access to the CPN tools described in Chap. 6 – or access to other Petri net tools.

At Aarhus University Vol. 1 is used as the material for a graduate course. The course has 10–12 lectures of 2 hours each. This is supplemented by 8–10 classes discussing the exercises. Moreover, the students make a small project in which they model an elevator system. The students are expected to use one third of their study time, during 4 months. The purpose of the course is to teach the students to use hierarchical CP-nets to model and simulate systems. Afterwards the students may follow another course, which presents the theory and application of the formal analysis methods.

Relationship to earlier definitions of CP-nets

The CP-nets presented in this book are, except for minor technical improvements of the notation, identical to the CP-nets presented in [58]. This means that the non-hierarchical CP-nets are similar to the CP-nets defined in [54] and the High-level Petri Nets (HL-nets) defined in [52].

In all these papers, CP-nets (and HL-nets) have two different representations. The *expression representation* uses arc expressions and guards, while the *function representation* uses linear functions mapping multi-sets into multi-sets. Moreover, there are formal translations between the two representations (in both directions). In [52] and [54] we used the expression representation to describe systems, while we used the function representation for all the different kinds of analysis. However, it has turned out that it is only necessary to turn to linear functions when we deal with invariants analysis, and this means that in [58] and this book we use the expression representation for all purposes – except for the calculation of invariants. This change is important for the practical use of

CP-nets, because it means that the function representation and the translations (which are a bit complicated mathematically) are no longer parts of the basic definition of CP-nets. Instead they are parts of the invariant method (which anyway demands considerable mathematical skills).

Acknowledgments

Many different people have contributed to the development of CP-nets and the CPN tools. Below some of the most important contributions are listed.

CP-nets were derived from Predicate/Transition Nets which were developed by Hartmann Genrich and Kurt Lautenbach. Many students and colleagues – in particular at Aarhus University – have influenced the development of CP-nets. Grzegorz Rozenberg has been a great support and inspiration for the book project (and for many other of my Petri net activities). The development of CP-nets has been supported by several grants from the Danish National Science Research Council.

The first version of occurrence graphs with equivalence classes was developed together with Peter Huber, Arne Møller Jensen and Leif Obel Jepsen. The hierarchy constructs were developed together with Peter Huber and Robert M. Shapiro. The idea to use an extension of Standard ML for the inscriptions of CP-nets is due to Jawahar Malhotra.

The CPN tools were designed together with Peter Huber and Robert M. Shapiro and implemented together with Greg Alonso, Ole Bach Andersen, Søren Christensen, Jane Eisenstein, Alan Epstein, Vino Gupta, Ivan Hajadi, Peter Huber, Alain Karsenty, Jawahar Malhotra, Roberta Norin, Valerio Pinci and Robert M. Shapiro. Hartmann Genrich participated in many parts of the CPN tool project and Bob Seltzer has been a continuous supporter of the project.

Meta Software has provided the financial support for the CPN tool project. So far more than 25 man-years have been used. The project is also supported by the Danish National Science Research Council, the Human Engineering Division of the Armstrong Aerospace Medical Research Laboratory at Wright-Patterson Air Force Base, and the Basic Research Group of the Technical Panel C3 of the US Department of Defense Joint Directors of Laboratories at the Naval Ocean Systems Center.

In addition to those mentioned above, a number of students and colleagues have read and commented on earlier versions of this book. In particular, I am grateful to Rikke Drewsen Andersen, Jonathan Billington, Allan Cheng, Geoff Cutts, Greg Findlow, Niels Damgaard Hansen, Jens Bæk Jørgensen, Horst Oberquelle, Laure Petrucci, Dan Simpson and Antti Valmari (scientific and linguistic assistance) – and to Karen Kjær Møller, Angelika Paysen and Andrew Ross (linguistic assistance).

Aarhus, Denmark
April 1992

Kurt Jensen

Table of Contents

Chapter 1

Informal Introduction to Coloured Petri Nets

High-level nets, such as Coloured Petri Nets (CP-nets) and Predicate/ Transition Nets (Pr/T-nets), are now in widespread use for many different practical purposes. A selection of references to papers describing industrial applications can be found in the bibliographical remarks of Chap. 7. The main reason for the great success of these kinds of net models is the fact that they have a *graphical representation* and a *well-defined semantics* allowing *formal analysis*. The step from low-level Petri nets to high-level nets can be compared to the step from assembly languages to modern programming languages with an elaborated type concept. In low-level nets there is only one kind of token and this means that the state of a place is described by an integer (and in many cases even by a boolean value). In high-level nets, each token can carry complex information or data (which, e.g., may describe the entire state of a process or a data base).

This chapter contains an informal introduction to CP-nets. Section 1.1 illustrates the structural and behavioural properties of low-level Petri nets, i.e., Petri nets without colours. Such nets are called Place/Transition Nets (PT-nets). They are introduced by means of a small example, which describes a set of processes sharing a common pool of resources. Analogously, Sect. 1.2 illustrates the structural and behavioural properties of CP-nets, and this is done by means of the same example. Section 1.3 contains a second example of a CP-net, describing a distributed data base. This example is used to illustrate some basic concepts of CP-nets – known as conflict, concurrency, and causal dependency. Section 1.4 introduces a language called CPN ML. This language is used for the net inscriptions, i.e., the text strings attached to the places, transitions and arcs of a CP-net. CPN ML is an extension of a well-known functional programming language, called Standard ML. Section 1.5 contains some simple guidelines, which may help the inexperienced modeller to construct his first CPN models. Analogously, Sect. 1.6 contains some guidelines telling how to draw more readable nets. Finally, Sect. 1.7 offers a brief description of some of the most important advantages of using CP-nets to describe and analyse systems.

1.1 Introduction to Place/Transition Nets

In this section we introduce the reader to ordinary Petri nets, i.e., Petri nets without colours. Such nets are called Place/Transition Nets (PT-nets).

We could start this section with a lengthy motivation for the qualities of PT-nets. Instead we shall simply take for granted that twenty-five years of theoretical work and practical experiences – documented in several thousand journal papers and research reports – have proved Petri nets to be one of the most useful languages for the modelling of systems containing concurrent processes. In Sect. 1.7 we shall give a number of different arguments for the use of CP-nets, and many of these will also be arguments in favour of PT-nets. Now, let us introduce PT-nets by means of a small example.

Assume that we have a set of processes, which share a common pool of resources. There are two different kinds of processes (called p-processes and q-processes) and three different kinds of resources (called r-resources, s-resources, and t-resources). The processes could be different computer programs (e.g., text editors and drawing programs) while the resources could be different facilities shared by the programs (e.g., tape drives, laser printers and plotters). Each process is cyclic and during the individual parts of its cycle, the process needs to have exclusive access to a varying amount of the resources. The demands of the processes are specified in Fig. 1.1. We can see that p-processes can be in four different states. In the upper state no resources are needed. In the second state two s-resources are needed, and so on. Analogously, q-processes have five different states, and for each of these states the required amount of resources is specified. Altogether, the system comprises two p-processes, three q-processes, one r-resource, three s-resources, and two t-resources. The two arcs

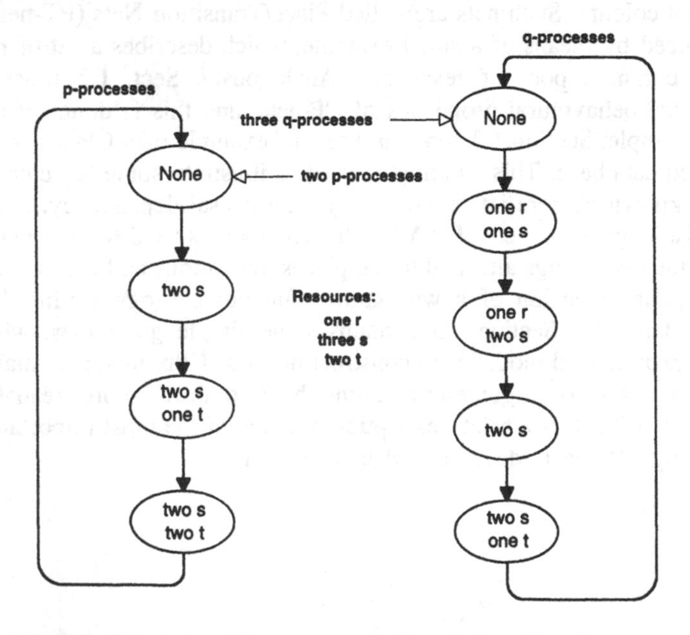

Fig. 1.1. The states of the processes in the resource allocation system

with white arrowheads indicate that all processes start in the two states at the top of the diagram.

In Fig. 1.1 we specified the demands of the processes by describing the possible **states**. Alternatively, we might have specified the demands by describing the possible **actions,** as shown in Fig. 1.2. The two arcs with white arrowheads indicate that all processes start by executing the two actions at the top of the diagram.

The specification in Fig. 1.1 is **state oriented** because it gives an explicit description of the possible states. The possible actions – which are needed to switch between the states – are only implicitly described (they can be deduced from the states). In contrast, the specification in Fig. 1.2 is **action oriented** because it gives an explicit description of the possible actions. The possible states – which exist between the actions – are only implicitly described (they can be deduced from the actions). It is easy to confirm that the two specifications are consistent (i.e., they do not contradict each other). In fact they are equivalent (i.e., the same amount of information can be deduced from them).

Figure 1.3 gives a Petri net specification of the resource allocation system. A Petri net is **state and action oriented** at the same time – because it gives an explicit description of both the states and the actions. This means that the user can determine freely whether – at a given moment of time – he wants to concentrate on states or on actions. The PT-net in Fig. 1.3 can be considered to be obtained by combining the state description of Fig. 1.1 with the action description of Fig. 1.2.

The states of the resource allocation system are indicated by means of ellipses (in this case circles), which are called **places**. There are twelve different places. Three of the places represent the three different kinds of resources (R, S and T).

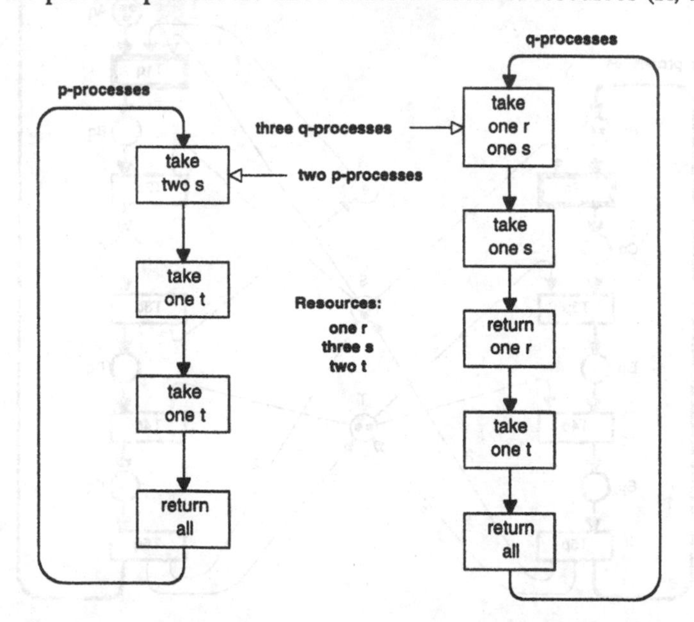

Fig. 1.2. The actions of the processes in the resource allocation system

Four of the places represent the possible states of the p-processes (Bp, Cp, Dp and Ep), while the five remaining places represent the states of the q-processes (Aq, Bq, Cq, Dq and Eq). The names of the places have no formal meaning – but they have an immense practical importance for the readability of a CP-net and they make it easy to refer to the individual places. The use of adequate mnemonic place names is strongly recommended. This is analogous to the use of well-chosen variable and procedure names in traditional programming.

Each place may contain a dynamically varying number of small black dots, which are called **tokens**. An arbitrary distribution of tokens on the places is called a **marking**. Figure 1.3 shows the initial distribution of tokens on the places. This is called the **initial marking** and it is usually denoted by M_0. The two tokens on Bp tell us that there are two p-processes and they both start in the state Bp. Analogously, there are three q-processes and they all start in state Aq. The tokens on the resource places R, S and T tell us the number of available resources (one r-resource, three s-resources and two t-resources).

The actions of the resource allocation system are indicated by means of rectangles, which are called **transitions**. There are nine different transitions. Four of the transitions represent the possible actions of the p-processes (T2p, T3p, T4p and T5p), while the five remaining transitions represent the actions of the q-processes (T1q, T2q, T3q, T4q and T5q). The names of the transitions have no formal meaning – but have a similar practical importance as the place names. The places and transitions of a PT-net are collectively referred to as the **nodes**.

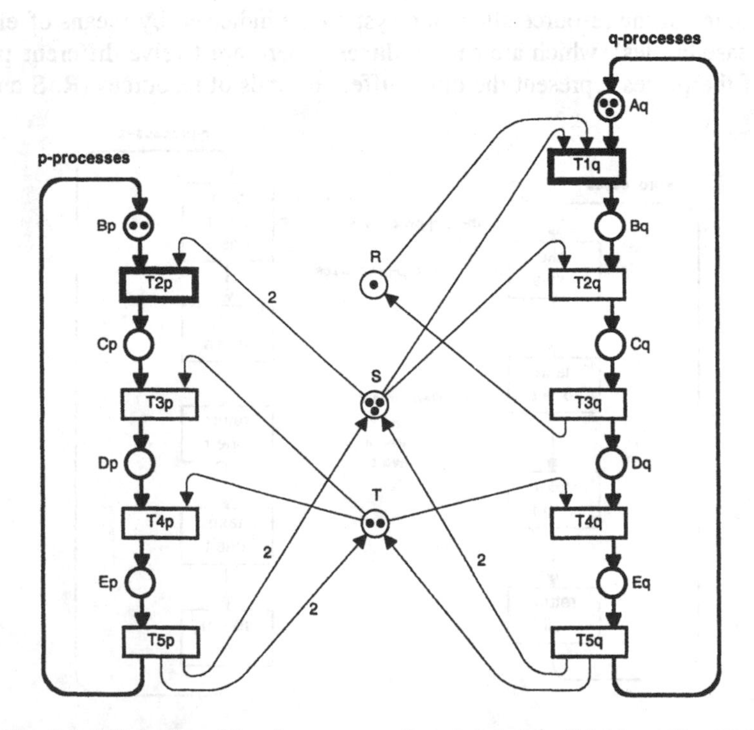

Fig. 1.3. PT-net describing the resource allocation system (initial marking M_0)

The PT-net also contains a set of directed arrows, which are called **arcs**. Each arc connects a place with a transition or a transition with a place – but never two nodes of the same kind. Each arc may have a positive integer attached to it (e.g., see the arc from place S to transition T2p). This integer is called an **arc expression**. By convention we omit arc expressions which are identical to 1.

A node x is called an **input node** of another node y, iff there exists a directed arc from x to y (we use the term "iff" as a shorthand for "if and only if"). Analogously, a node x is called an **output node** of another node y, iff there exists a directed arc from y to x. We shall also talk about **input places, output places, input transitions, output transitions, input arcs** and **output arcs**. As an example, transition T2p has two input arcs connecting it with the input places Bp and S, and it has one output arc connecting it with the output place Cp. Analogously, place S has two input arcs connecting it with the input transitions T5p and T5q, and it has three output arcs connecting it with the output transitions T2p, T1q and T2q.

In Fig. 1.3 the left-hand part and the right-hand part of the net (describing how processes change between different states) are drawn with thick lines. This distinguishes them from the rest of the net (describing how resources are reserved and released). However, it should be stressed that such graphical conventions have no formal meaning. Their only purpose is to make the net more readable for human beings – and hence they are analogous to the use of a consistent indentation strategy in ordinary programming languages.

Above we have discussed the syntax of PT-nets. Now let us consider the semantics, i.e., the behaviour. A PT-net can be considered to be a game board where the tokens are markers (which are only allowed to be positioned on the places). Each transition represents a potential move in the "Petri net game". A move is possible iff each input place of the transition contains at least the number of tokens prescribed by the arc expression of the corresponding input arc. We then say that the transition is **enabled**. In the initial marking of Fig. 1.3 transition T2p is enabled because there is at least one token on Bp and two tokens on S. Analogously, transition T1q is enabled because there is at least one token on each of the places Aq, R and S. All other transitions are **disabled** because there are too few tokens on some of their input places. In Fig. 1.3 we have used a different line thickness to indicate that T2p and T1q are enabled.

When a transition is enabled the corresponding move *may* take place. If this happens we say that the transition **occurs**. The effect of an occurrence is that tokens are removed from the input places and added to the output places. The number of removed/added tokens is specified by the arc expression of the corresponding input/output arc. As an example the occurrence of transition T2p transforms the initial marking M_0 (Fig. 1.3) into the marking M_1 which is shown in Fig. 1.4. We often think of this transformation as if a token is moved from Bp to Cp, while two tokens are removed from S. However, it should be stressed that the Petri net formalism does not talk about tokens being moved from input places to output places. Instead tokens are removed from the input places, and completely new tokens are added to the output places. This means that there is no closer relationship between the token removed at Bp and the token added at Cp than there is between the tokens removed from S and the token added to Cp. It is only because of our intuitive interpretation of the model that we sometimes think of the token as being moved from Bp to Cp (representing that a p-process changes its state from B to C).

Analogously, the occurrence of transition T1q transforms M_0 (Fig. 1.3) into the marking M_2 which is shown in Fig. 1.5. We say that each of the two markings M_1 and M_2 are **directly reachable** from M_0 – by the occurrence of T2p and T1q, respectively.

In M_1 (Fig. 1.4) each of the transitions T3p and T1q are enabled (as indicated by the additional line thickness). An occurrence of T3p yields a marking which is directly reachable from M_1 and **reachable** from M_0 (by the occurrence of T2p followed by T3p).

In M_0 (Fig. 1.3) we have seen that each of the transitions T2p and T1q are enabled. This means that each of them may occur. Moreover, there are so many tokens on the common input place S (shared by T2p and T1q) that each of the two transitions can get their "private" tokens – without having to "share" them with the other transition. Thus we say that T2p and T1q are **concurrently enabled** in M_0. This means that the two transitions may occur "at the same time" or "in parallel". We also say that the **step** $S_1 = \{T2p, T1q\}$ is enabled in M_0.

A transition may even occur **concurrently to itself**. If we in the initial marking of the resource allocation system add two extra e-tokens to S (so that we get five s-resources), each of the following steps would be enabled in the modified marking, M*:

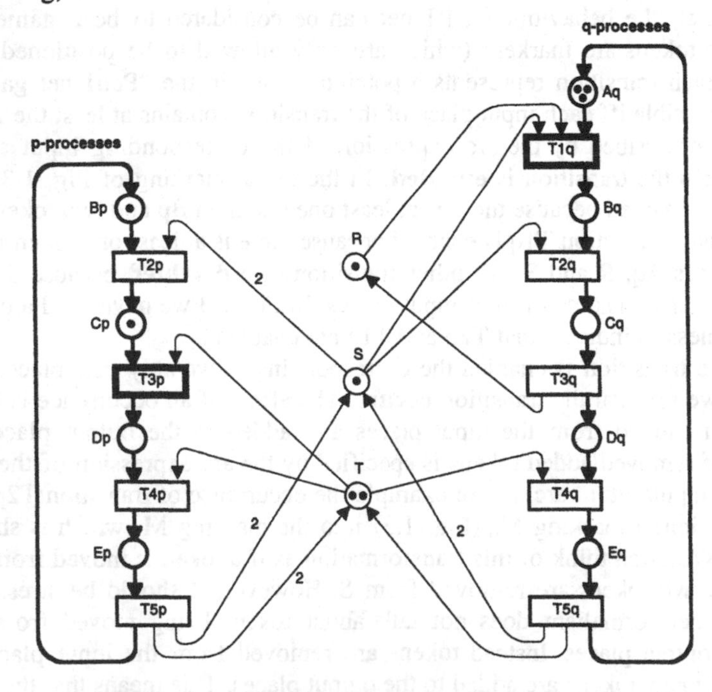

Fig. 1.4. The marking M_1 (reachable from M_0 by T2p)

S_2 = {T2p, T2p, T1q}
S_3 = {T2p, T2p}
S_4 = {T2p, T1q}
S_5 = {T2p}
S_6 = {T1q}.

In general, each step is a **multi-set** – over the set of all transitions. A multi-set is analogous to a set, except that it may contain multiple appearances of the same element. A formal definition of multi-sets will be given in Chap. 2. The multi-set {T2p, T2p, T1q} is said to contain two **appearances** of T2p, one appearance of T1q, and no appearances of the other transitions. We usually denote multi-sets by sums where each element has a **coefficient** telling how many times it appears. Using this notation the enabled steps of M* are denoted as shown below. As an example, the first line should be read: S_2 contains two appearances of T2p plus one appearance of T1q. By convention we omit elements with zero coefficients, and we insert ` between each coefficient and element. We shall return to the details of this notation in Sect. 2.1:

S_2 = 2`T2p + 1`T1q
S_3 = 2`T2p
S_4 = 1`T2p + 1`T1q
S_5 = 1`T2p
S_6 = 1`T1q.

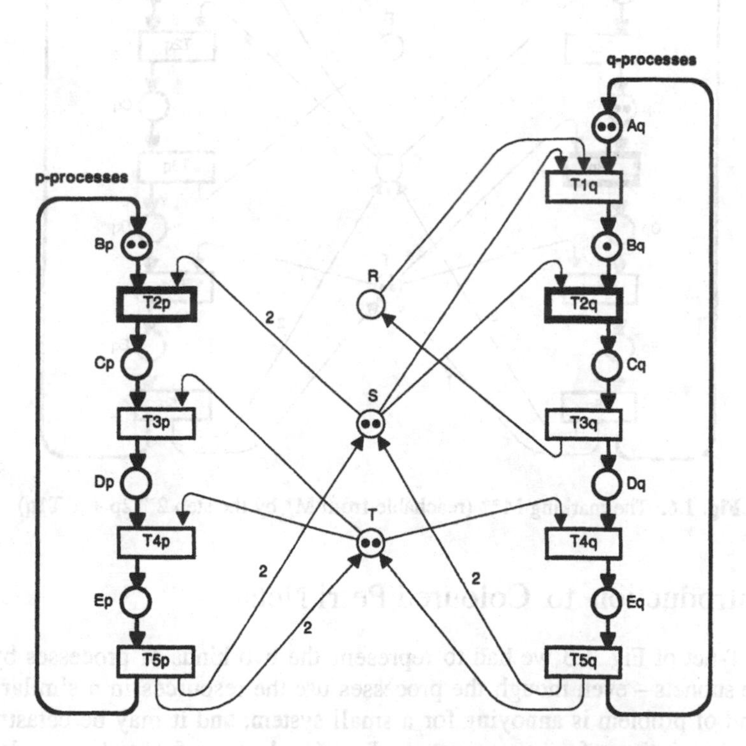

Fig. 1.5. The marking M_2 (reachable from M_0 by T1q)

It should be stressed that two transitions are only concurrently enabled if they are **independent** in the sense that they can operate on disjoint sets of tokens. In M_2 (Fig. 1.5) both T2p and T2q are (individually) enabled, but they are *not* concurrently enabled, because they would have to "share" one of the tokens on S. When a step with several transitions occurs, the effect is the sum of the effects of the individual transitions. As an example, S_2 transforms the marking M* into the marking M** which is shown in Fig. 1.6. Notice that neither M* nor M** is reachable from M_0.

The step S_2 contains each of the other steps S_3, S_4, S_5 and S_6 as a "sub-multi-set". Moreover, it is easy to see that the enabling of a step implies that each smaller step also is enabled. This is a particular case of a more general property of Petri nets. Whenever an enabled step has a size which is larger than one, the step can (in any thinkable way) be divided into two or more steps – which then are known to be able to occur after each other (in any order) and together have the same total effect as the original step.

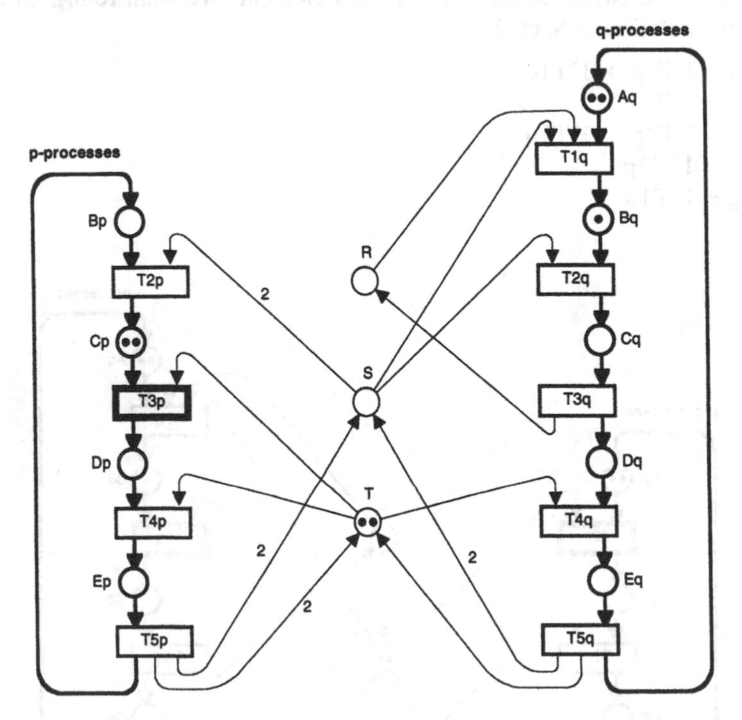

Fig. 1.6. The marking M** (reachable from M* by the step 2\`T2p + 1\`T1q)

1.2 Introduction to Coloured Petri Nets

In the PT-net of Fig. 1.3 we had to represent the two kinds of processes by two separate subnets – even though the processes use the resources in a similar way. This kind of problem is annoying for a small system, and it may be catastrophic for the description of a large system. Imagine how awful and unreadable a

PT-net will be, if the resource allocation system contains 20 different kinds of processes and resources.

Unfortunately, this problem is typical for many practical applications of Petri nets. Real-world systems often contain many parts which are similar, but not identical. Using PT-nets, these parts must be represented by disjoint subnets with a nearly identical structure. This means that the total PT-net becomes very large. Moreover, it becomes difficult to see the similarities (and differences) between the individual subnets representing similar parts.

The practical use of PT-nets to describe real-world systems has clearly demonstrated a need for more powerful net types, to describe complex systems in a manageable way. The development of high-level Petri nets constitutes a very significant improvement in this respect. CP-nets belong to the class of high-level nets. Other important representatives of high-level nets will be mentioned in the bibliographic remarks. The CP-net in Fig. 1.7 (to be explained below) describes the same system as the PT-net in Fig. 1.3 – but now in a more compact form, which stresses the similarities (and differences) between the two kinds of processes. To indicate how additional information or data can be handled in a CP-net, without making the net structure more complex, we now also count how many full cycles each process has completed.

The more compact representation has been achieved by equipping each token with an attached data value – called the **token colour**. The data value may be of arbitrarily complex type (e.g., a record where the first field is a real, the second a text string, while the third is a list of integer pairs). For a given place all tokens must have token colours that belong to a specified type. This type is called the **colour set** of the place. By convention we write colour sets in italics, and from Fig. 1.7 it can be seen that the places A–E have the type P as colour set, while the places R–T have the type E as colour set.

The use of colour sets in CP-nets is totally analogous to the use of types in programming languages. Colour sets determine the possible values of tokens (analogously to the way in which types determine the possible values of variables and expressions). For historical reasons we talk about "coloured tokens" which can be distinguished from each other – in contrast to the "plain tokens" of PT-nets. However, we shall often also allow ourselves to talk about values and types (instead of colours and colour sets).

From the **declarations** (in the dashed box at the upper left corner of Fig. 1.7) it can be seen what the colour sets are. In this book we use a language called **CPN ML** for the declarations. This language is used by the CPN tools described in Chap. 6 (and it uses American spelling, in contrast to the rest of this book which uses British spelling). We shall return to CPN ML in Sect. 1.4. The declarations of the colour sets tell us that each token on A–E has a token colour which is a pair (because the colour set P is declared to be the cartesian product of two other colour sets U and I). The first element of the pair is an element of U and thus it is either p or q (because the colour set U is declared to be an enumeration type with these two elements). The second element is an integer (because the colour set I is declared by means of the CPN ML standard type int, which contains all integers in an implementation-dependent interval). Intuitively, the first element of a token tells whether the token represents a p-process or a q-process, while the second element tells how many full cycles the process has completed. It can also be seen that all the tokens on R–T have the same token colour (e is the only element of E). Intuitively, this means that these tokens carry no information – apart from their presence/absence at a place.

A distribution of tokens (on the places) is called a **marking.** The **initial marking** is determined by evaluating the **initialization expressions**, i.e., the underlined expressions next to the places. In the initial marking there are three (q,0)-tokens on A and two (p,0)-tokens on B, while C, D and E have no tokens (by convention we omit initialization expressions which evaluate to the empty multi-set). Moreover, R has one e-token, S has three e-tokens and T has two e-tokens. The marking of each place is a **multi-set** over the colour set attached to the place. We need multi-sets to allow two or more tokens to have identical token colours. If we only worked with sets it would be impossible, for example, to have three (q,0)-tokens in the initial marking of A. As a shorthand we shall also allow initialization expressions which evaluate to a single colour c, and interpret this as if the value was 1`c (i.e., the multi-set which contains one appear-

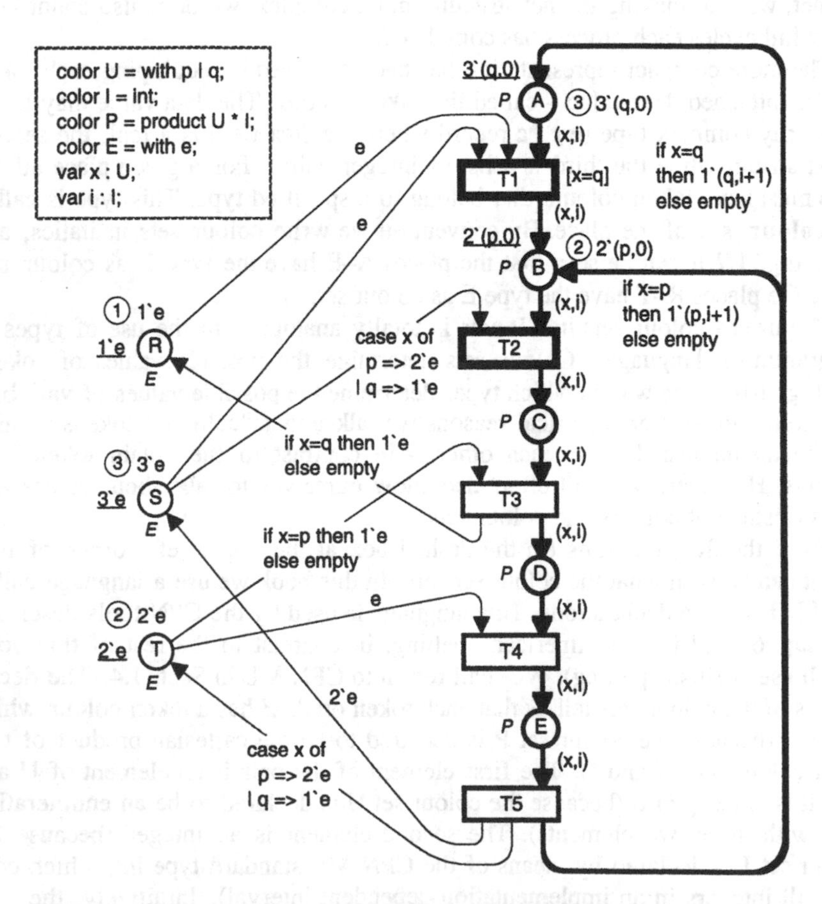

Fig. 1.7. CP-net describing the resource allocation system (initial marking M_0)

ance of c, and nothing else). Thus, for the place R, we could have written "e" instead of "1`e".

In the PT-nets we indicated the **current marking** by putting small dots inside the individual places. However, this is not practical for CP-nets – because in addition to the number of tokens we also have to represent the token colours (which may be arbitrarily complex data values). Thus we shall instead represent the current marking of a given place by means of a small circle (with an integer saying how many tokens there are) and a text string next to the circle (with a multi-set saying what the individual token colours are, and which coefficients they have). By convention we omit the circle and the text string for places which have no tokens. The notation with the circles and text strings is adopted from the CPN simulator described in Chap. 6. In the simulator it is possible to make the individual text strings (which may be very large) visible/invisible by double-clicking on the corresponding circle. In this way it is possible to display complex markings without overloading the diagram with too many large text strings. In Fig. 1.7 the current marking is identical to the initial marking, and this means that the small circles and their text strings contain the same information as the initialization expressions.

Attaching a colour to each token and a colour set to each place allows us to use fewer places than would be needed in a PT-net. As an example, we only have one B place in Fig. 1.7, while in Fig. 1.3 we had both a Bp and a Bq place. Intuitively, the introduction of colours has allowed us to fold Bp and Bq into a single place – without losing the ability to distinguish between p-processes and q-processes. This may not seem too impressive – but imagine a resource allocation system where we have many different kinds of processes and many different kinds of resources, or the system where in the PT-net we also want to keep track of the number of full cycles completed by the individual processes. In these cases we would have a much more significant gain, with respect to the number of places and transitions.

Allowing more complex markers also means that the moves in the "Petri net game" become more complex. The token colours can be inspected by the transitions, which means that the enabling of a transition may depend upon the colours of its input tokens. It also means that the colours of the input tokens may determine the colours of the output tokens produced, i.e., the effect of the transition. To describe this more complex situation, we need more elaborate **arc expressions**. It is no longer sufficient to have an integer specifying the number of tokens which are added/removed. Instead we need arc expressions which specify a collection of tokens – each with a well-defined token colour. To do this we use arc expressions which evaluate to **multi-sets**. As a shorthand, we shall also allow arc expressions which evaluate to a single colour c, and interpret this as if the value were 1`c.

Above we have discussed how the introduction of colours allowed us to replace a large number of PT-net places with a single CP-net place (without losing precision in our information about the system). However, this only solves half of the problem (and explains half of the differences between Fig. 1.3 and Fig. 1.7). To get a compact Petri net we not only need fewer places, we also need fewer transitions. Thus we will now investigate how we can replace a number of PT-net transitions with a single CP-net transition. This means that we need to develop a method by which each CP-net transition can occur in many different ways – representing the different actions modelled by all the corresponding PT-net transitions. These actions are closely related, but slightly different from each other. This is analogous to the execution of procedures. A procedure represents a number of executions which are closely related (because they use the same code) but slightly different from each other (because the actual parameters may vary from call to call). The situation is also analogous to the calculation of expressions containing variables. Again each expression represents a number of calculations which are closely related (because they all use the same expression) but slightly different from each other (because the variables of the expression may have different values from evaluation to evaluation). The above analogy between expression evaluation and CP-net transitions actually brings us right to the answer to our problem. To get transitions which can represent many slightly different actions, we allow the arc expressions surrounding a given transition to contain a number of variables. These variables can be bound to different values and this means that the expressions evaluate to different values (i.e., different multi-sets of token colours).

Now let us consider the arc expressions in Fig. 1.7. Around transition T2 we have three arc expressions: "(x,i)" appears twice (on the input arc from B and on the output arc to C) while "case x of p=>2`e | q=>1`e" appears once (on the input arc from S). Together these three arc expressions have two variables, x and i, and from the declarations it can be seen that x has type U while i has type I. At a first glance one might also think that e, p and q are variables, but from the declarations it can be seen that this is not the case: e is an element of the colour set E, while p and q are elements of U. This means that they are constants. Intuitively, the three arc expressions tell us that an occurrence of T2 moves a token from B to C – without changing the colour (because the two arc expressions are identical). Moreover, the occurrence removes a multi-set of tokens from S. This multi-set is determined by evaluating the corresponding arc expression. As it can be seen, the multi-set depends upon the kind of process which is involved. A p-process needs two s-resources to go from B to C (and thus it removes two e-tokens from S), while a q-process only needs one s-resource to go from B to C (and thus it removes only one e-token from S).

Now let us be a little more precise, and explain in detail how the enabling and occurrence of CP-net transitions are calculated. The transition T2 has two variables (x and i), and before we can consider an occurrence of the transition these variables have to be bound to colours of the corresponding types (i.e., elements of the colour sets U and I). This can be done in many different ways. One possibility is to bind x to p and i to zero: then we get the **binding** b_1 = <x=p,i=0>. Another possibility is to bind x to q and i to 37: then we get the binding b_2 = <x=q,i=37>.

For each binding we can check whether the transition, with that binding, is **enabled** (in the current marking). For the binding b_1 the two input arc expressions evaluate to (p,0) and 2`e, respectively. Thus we conclude that b_1 is enabled in the initial marking – because each of the input places contains at least the tokens to which the corresponding arc expression evaluates (one (p,0)-token on B and two e-tokens on S). For the binding b_2 the two arc expressions evaluate to (q,37) and e. Thus we conclude that b_2 is *not* enabled (there is no (q,37)-token on B). A transition can occur in as many ways as we can bind the variables that appear in the surrounding arc expressions (and in the guard – introduced below). However, for a given marking, it is usually only a few of these bindings that are enabled.

When a transition is enabled (for a certain binding) it may **occur**, and it then removes tokens from its input places and adds tokens to its output places. The number of removed/added tokens and the colours of these tokens are determined by the value of the corresponding arc expressions (evaluated with respect to the binding in question). A pair (t,b) where t is a transition and b a binding for t is called a **binding element.** The binding element (T2,b_1) is enabled in the initial marking M_0 and it transforms M_0 into the marking M_1 which is shown in Fig. 1.8. Analogously, we conclude that the binding element (T1,<x=q,i=0>) is enabled in M_0 and that it transforms M_0 into the marking M_2 which is shown in Fig. 1.9. We say that each of the markings M_1 and M_2 is **directly reachable** from M_0. The binding element (T2,b_2) is *not* enabled in M_0 and thus it cannot occur.

Notice that the two CP-net markings M_1 (Fig. 1.8) and M_2 (Fig. 1.9), which can be reached from the CP-net marking M_0 (Fig. 1.7) by the occurrence of (T2,<x=p,i=0>) and (T1,<x=q,i=0>), are analogous to the two PT-net markings M_1 (Fig. 1.4) and M_2 (Fig. 1.5) which can be reached from the PT-net marking M_0 (Fig. 1.3) by the occurrence of T2p and T1q.

Next let us look at transition T1, which in addition to the arc expressions has a **guard**: x=q. The guard is a boolean expression (i.e., an expression which evaluates to either true or false) and it may have variables in exactly the same way that the arc expressions have. The purpose of the guard is to define an additional constraint which must be fulfilled before the transition is enabled. In this case the guard tells us that it is only tokens representing q-processes which can move from A to B (because the guard for all bindings <x=p,...> evaluates to false and thus prevents enabling). In Fig. 1.7 we could have omitted the guard, since from the initial marking and a simple analysis of the net it can be concluded that only q-processes can be in state A. However, in more complex situations guards turn out to be extremely useful. By convention we omit guards which always evaluate to true (in a similar way that we omit initialization expressions which evaluate to the empty multi-set). We also allow a guard to be a *list* of boolean expressions [$Bexpr_1$, $Bexpr_2$, ...,$Bexpr_n$], and this is a shorthand for the boolean expression $Bexpr_1 \wedge Bexpr_2 \wedge \ldots \wedge Bexpr_n$. A list [Bexpr] with a single boolean expression is equivalent to Bexpr itself, and this means that we can always enclose guards in square brackets, as indicated by [x=q]. This is recommended because it makes it easier to distinguish guards from the other kinds of net inscriptions, e.g., arc expressions and colour set inscriptions.

To illustrate different possibilities the CP-net of Fig. 1.7 contains both case-expressions and if-expressions. However, it should be obvious that each case-expression can be rewritten as an equivalent if-expression, and vice versa. As an example, "case x of p=>2`e | q=>1`e" is equivalent to "if x=p then 2`e else 1`e".

Finally, let us take a closer look at transition T5. This transition moves a token from E to either A or B (p-processes go to B, while q-processes go to A). Simultaneously the transition updates the cycle counter *i*. Notice that different bindings for a transition may not only result in different token colours but also in different *numbers* of tokens. In particular this may mean that the multi-set of tokens which are added/removed, for a given binding, may be empty, as illustrated by the two thick output arcs of T5. We have positioned the first segments of the two arcs on top of each other to illustrate the close relationship between them. However, it should be stressed that this has no formal meaning. The only purpose is to make the drawing more readable for human beings.

Analogously to PT-nets, two transitions in a CP-net can be **concurrently enabled** if there exist bindings for the variables (in the guard and the surrounding arc expressions), such that the transitions use disjoint sets of tokens. As an example, the marking M_0 (Fig. 1.7) has an enabled **step** which looks as follows:

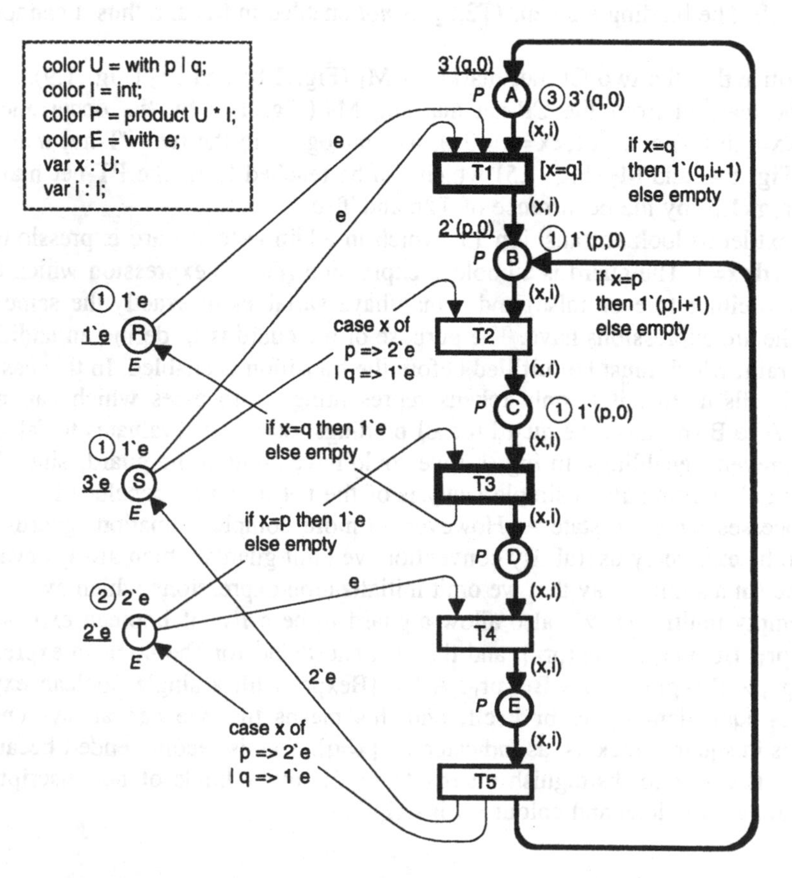

Fig. 1.8. The marking M_1 (reachable from M_0 by (T2,<x=p,i=0>))

S_1 = 1\`(T1,<x=q,i=0>) + 1\`(T2,<x=p,i=0>).

This means that we can have a step where both (T1,<x=q,i=0>) and (T2,<x=p,i=0>) occur. The effect of the step is the *sum* of the effects of the individual binding elements. This means that the occurrence of S_1 moves a (q,0)-token from A to B, moves a (p,0)-token from B to C, and removes a single e-token from R and three e-tokens from S. Notice that the effect of the step S_1, by definition, is the same as when the two binding elements occur after each other in arbitrary order. This is a general property of Petri nets. Whenever an enabled step contains more than one binding element, it can (in any thinkable way) be divided into two or more steps, which then are known to be able to occur after each other (in any thinkable order) and together have the same total effect as the original step.

When the same variable name appears more than once, in the guard/arc expressions of a *single* transition, we only have one variable (with multiple appearances). Each binding of the transition specifies a colour for the variable – and this colour is used for all the appearances. However, it should be noted that the appearances of x around T1 are totally independent of the appearances of x

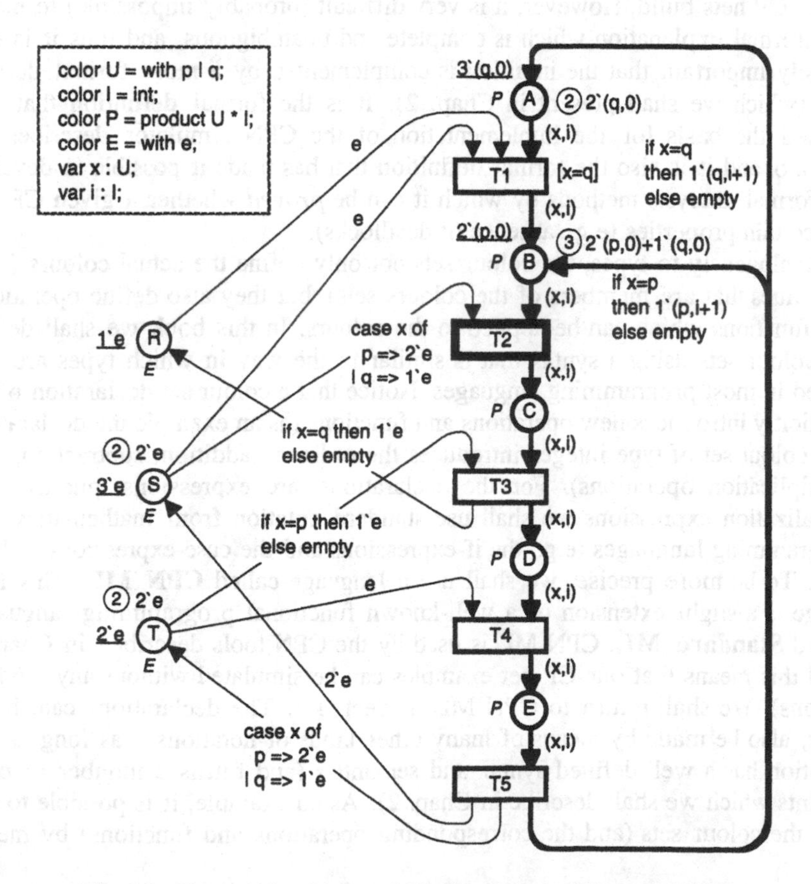

Fig. 1.9. The marking M_2 (reachable from M_0 by (T1,<x=q,i=0>))

around T2 – in the sense that the two sets of appearances in the same step can be bound to different colour values (as illustrated by S_1).

It is also possible for a transition to be **concurrently enabled with itself**. As an example, in the marking M_2 (Fig. 1.9) we could add two extra e-tokens to the place S (so that it gets four e-tokens). Then the following two steps would be enabled:

S_2 = 1`(T2,<x=p,i=0>) + 1`(T2,<x=q,i=0>)
S_3 = 2`(T2,<x=p,i=0>).

An occurrence of S_2 moves a (p,0)-token and a (q,0)-token from B to C, and it removes three e-tokens from S. An occurrence of S_3 moves two (p,0)-tokens from B to C, and it removes four e-tokens from S. When a step has two different bindings of the same transition (as illustrated by S_2) it should be noted that these bindings are totally independent – in the sense that they may specify different colours for the same variable (as illustrated by the variable x in S_2). Each of the two colours applies to all appearances of x around T2.

The above informal explanation of the enabling and occurrence rules tells us how to understand the behaviour of a CP-net, and it explains the intuition on which CP-nets build. However, it is very difficult (probably impossible) to make an informal explanation which is complete and unambiguous, and thus it is extremely important that the intuition is complemented by a more formal definition (which we shall present in Chap. 2). It is the formal definition that has formed the basis for the implementation of the CPN simulator described in Chap. 6, and it is also the formal definition that has made it possible to develop the formal analysis methods by which it can be *proved* whether a given CP-net has certain properties (e.g., absence of deadlocks).

Analogously to types, the colour sets not only define the actual colours (i.e., the values that are members of the colours sets), but they also define operations and functions which can be applied to the colours. In this book we shall define the colour sets using a syntax that is similar to the way in which types are declared in most programming languages. Notice that a colour set declaration often implicitly introduces new operations and functions (as an example the declaration of a colour set of type integer introduces the ordinary addition, subtraction, and multiplication operations). For the declarations, arc expressions, guards, and initialization expressions we shall use standard notation from mathematics and programming languages (e.g., the if-expressions and the case-expressions in Fig. 1.7). To be more precise, we shall use a language called **CPN ML**. This language is a slight extension of a well-known functional programming language, called **Standard ML**. CPN ML is used by the CPN tools described in Chap. 6 (and this means that our CP-net examples can be simulated without any modifications). We shall return to CPN ML in Sect. 1.4. The declarations can, however, also be made by means of many other kinds of notations – as long as the notation has a well-defined syntax and semantics (and fulfils a number of constraints which we shall describe in Chap. 2). As an example, it is possible to define the colour sets (and the corresponding operations and functions) by means of a many-sorted sigma algebra – in a way similar to that which is known from the theory of abstract data types.

Above we have seen that a CP-net consists of three different parts: the **net structure** (i.e., the places, transitions and arcs), the **declarations** and the **net inscriptions** (i.e., the various text strings which are attached to the elements of the net structure). As we shall see later, there is a well-defined set of rules saying what the different kinds of net inscriptions contain and how they influence the semantics. In Fig. 1.7 places have four different kinds of inscriptions, specifying the names, colour sets, initialization expressions and current markings. Transitions have two kinds of inscriptions, specifying the names and guards. Finally, arcs only have one kind of inscription, specifying the arc expressions. All net inscriptions are positioned next to the corresponding net element – and to make it easy to distinguish between them we write names in plain text, colour sets in italics, while initialization expressions are underlined and guards are enclosed in square brackets. Names have no formal meaning. They only serve as a means of identification that makes it possible for people and computer systems to refer to the individual places and transitions. Names can be omitted and one can use the same name for several nodes (although this may create confusion). As explained later in this book, a CP-net may have several other kinds of inscriptions (e.g., describing hierarchical relationships and time delays).

The complexity of a CP-net description is distributed among the net structure, the declarations and the net inscriptions – and this can be done in many different ways, as illustrated by the three CP-nets in Figs. 1.10–1.12. Each of these CP-nets is behaviourally equivalent to the CP-net in Fig. 1.7. Notice that we usually show a **CPN diagram**, such as Figs. 1.10–1.12, *without* a current marking and an enabling indication – but with the initialization expressions (from which the initial marking can be calculated).

In Fig. 1.10 each arc to or from a resource place has a very simple arc expression which either is a constant multi-set or of the form $F_i(x)$ where the function F_i is declared in the declaration part. This means that, compared with Fig. 1.7, we have moved information from the net inscriptions to the declarations. This is useful, in particular, when a large/complex arc expression is used for several different arcs.

In Fig. 1.11 we have represented all resources by means of a single place *Res*, having V={r, s, t} as colour set. The use of this colour set means that we still are able to distinguish between tokens representing r-resources, s-resources and t-resources – although they now all reside on the same place. The + operator in the arc expressions denotes addition of multi-sets. As an example, 2\`s+1\`t is the multi-set that contains two appearances of s and one appearance of t. In this case, compared with Fig. 1.7, we have moved information from the net structure to the net inscriptions.

In Fig. 1.12 we have folded the five transitions T1–T5 into a single transition *Move to Next State*, and the five places A–E into a single place, called *State*. The new place has the cartesian product U×S×I as the set of possible token colours, and this means that each token has attached information which is a triple (x,y,i). The first component $x \in U$ tells us whether the token represents a p-process or a q-process, the second component $y \in S = \{a,b,c,d,e\}$ tells us which of the five states the process is currently in, and the third component $i \in I$ tells us how many full cycles the process has completed. The function *Succ* gives this successor of each element in S (in a cyclic way). The function *Next* calculates the token colour of the new token to be put on *State*. Finally the functions *Reserve* and *Release* calculate the multi-sets of reserved and released resources ("|" separates the individual clauses of the case-expressions; "_" is analogous to an otherwise-clause, i.e., it takes care of all those cases which are not explicitly mentioned). In this case, compared with Fig. 1.7, we have moved information from the net structure to the declarations.

The CPN ML language allows the modeller to use a number of predefined functions, and this means that in Fig. 1.12 we could have used a predefined rotation function instead of declaring our own successor function. Then we would add "declare rot" to the end of the declaration of the colour set S – this would instruct the CPN ML compiler to include the declaration of a standard rotation

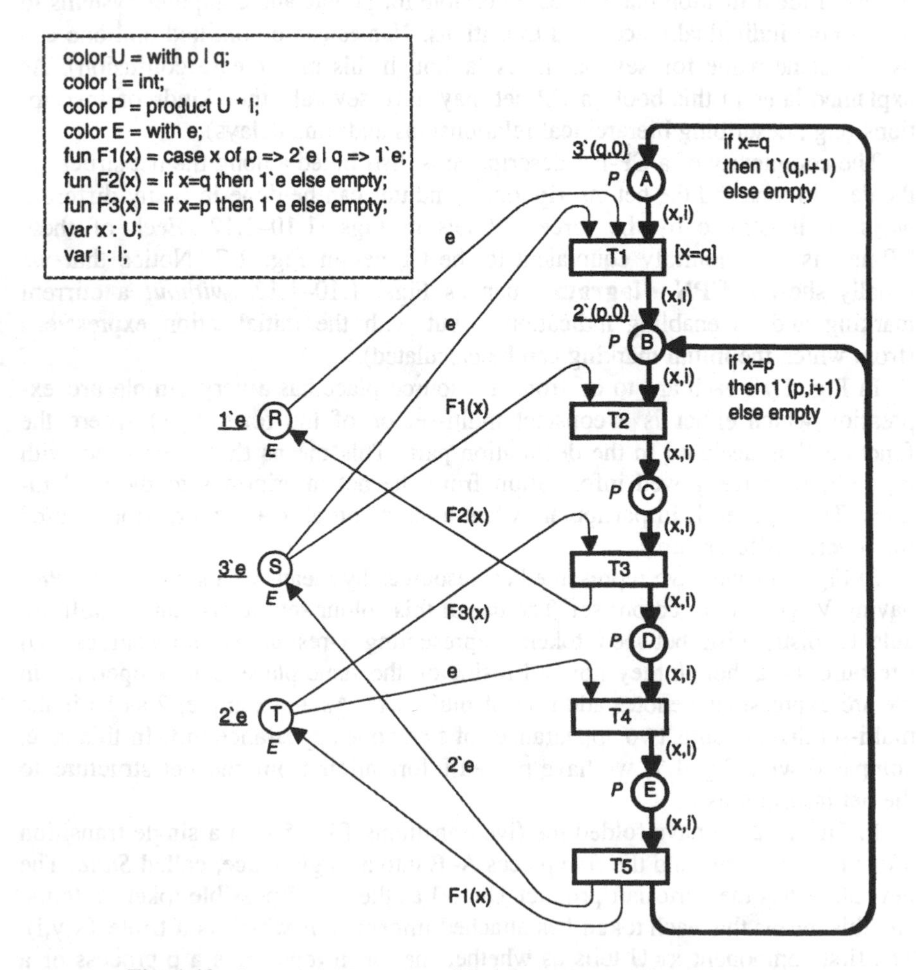

Fig. 1.10. A second CP-net describing the resource allocation system

function, rot'S. Moreover, we would have written "rot'S(y,1)" instead of "Succ(y)". A specification of the predefined CPN ML functions is outside the scope of this book, but it can (of course) be found in the CPN ML manual.

For nearly all purposes the CP-net in Fig. 1.12 is too compact, in the sense that most of the information is present in the declarations and the net inscriptions, while the net structure is very meagre. Usually, it is a good idea to try to distribute the information between all three parts of a CP-net, i.e., between the net structure, the declarations and the net inscriptions. Many modellers start with a relatively large net structure which is then later made more condensed – by "folding" related transitions into a single transition, and related places into a single place (in a way which is analogous to the transformation from Fig. 1.3 to Fig. 1.7). Such transformations can be specified by formal transformation rules. Then it is possible to create computerized Petri net editors where the modeller specifies the transformations to be performed, while the editor checks the validity and performs the detailed calculations.

As we shall see in Sect. 2.4, it is always possible to translate a PT-net to a CP-net (and as indicated above this can be done in a number of different ways). This means that we could start each modelling project by creating a PT-net, which we then transform into a CP-net. However, this is *not at all* the intention

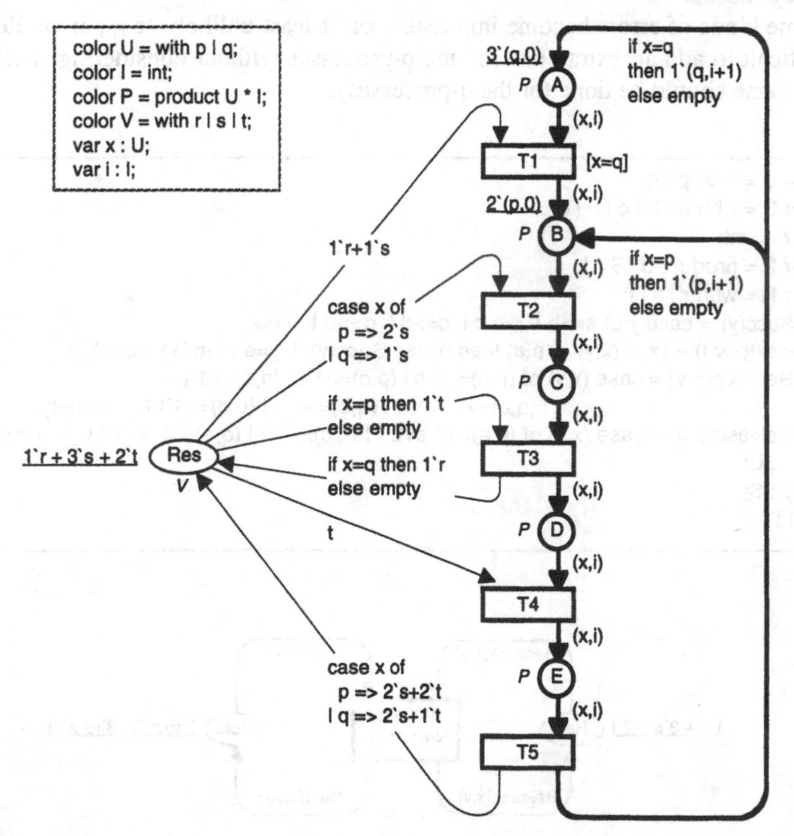

Fig. 1.11. A third CP-net describing the resource allocation system

behind CP-nets. Instead CP-nets form a language in its own right, and this means that systems are modelled and analysed directly in terms of CP-nets – without thinking of PT-nets and without translating them into PT-nets. The benefits which we achieve by using CP-nets, instead of PT-nets, are very much the same as those achieved by using high-level programming languages instead of assembly languages:

- Description and analysis become more compact and manageable (because the complexity is divided between the net structure, the declarations and the net inscriptions).
- It becomes possible to describe simple data manipulations in a much more direct way (by using the arc expressions instead of a complex set of places, transitions and arcs).
- It becomes easier to see the similarities and differences between similar system parts (because they are represented by the same subnet).
- The description is more redundant and this means that there will be less errors (because errors can be found by noticing inconsistencies, e.g., between the type of an arc expression and the colour set of the corresponding place). This is particularly useful when we have a computer tool to perform the consistency checks.
- Some kinds of errors become impossible or at least unlikely (e.g., it would be difficult to add an extra state for the p-processes without considering whether the same should be done for the q-processes).

```
color U = with p | q;
color S = with a | b | c | d | e;
color I = int;
color P = product U *S * I;
color R = with r | s | t;
fun Succ(y) = case y of a=>b | b=>c | c=>d | d=>e | e=>a;
fun Next(x,y,i) = (x, if (x,y) = (p,e) then b else Succ(y), if y=e then i+1 else i);
fun Reserve(x,y) = case (x,y) of (p,b)=>2`s | (p,c)=>1`t | (p,d)=>1`t
                                  | (q,a)=>1`r+1`s | (q,b)=>1`s | (q,d)=>1`t | _=>empty;
fun Release(x,y) = case (x,y) of (p,e)=>2`s+2`t | (q,c)=>1`r | (q,e)=>2`s+1`t | _=>empty;
var x : U;
var y : S;
var i : I;
```

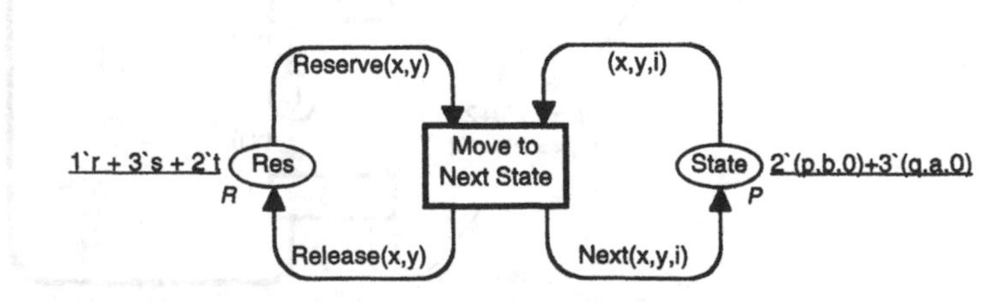

Fig. 1.12. A fourth CP-net describing the resource allocation system

- It is possible to create hierarchical descriptions, i.e., structure a large description as a set of smaller CP-nets with a well-defined relationship. Hierarchical CP-nets have not been mentioned above. They will be introduced in Chap. 3.

Before closing this section, let us again consider the resource allocation system in Fig. 1.7. It can easily be proved that this system has no deadlock (i.e., no reachable marking in which no binding element is enabled). This can be proved in several different ways, for example by means of place invariants or by means of occurrence graphs (two of the formal analysis methods which we shall consider in Chap. 5 and Vol. 2). Now let us, in the initial marking, add an extra s-resource (i.e., an extra e-token on S). This cannot lead to a deadlock, because deadlocks appear when we have too few resource tokens, and thus an extra resource token cannot cause a deadlock.

Is the argument above convincing? At a first glance: yes! However, the argument is *wrong*. Adding the extra s-resource actually means that we can reach a deadlock. This can be seen by letting the two p-processes advance from state B to state D, while the q-processes remain in state A.

Hopefully, this small example demonstrates that informal arguments about behavioural properties are dangerous – and this is one of our motivations for the development of the more formal analysis methods described in Chap. 5 and Vol. 2.

1.3 Example of CP-nets: Distributed Data Base

This section contains a second example of CP-nets. We describe a very simple distributed data base with n different sites (n is a positive integer, which is assumed to be greater than or equal to 3). Each site contains a copy of all data and this copy is handled by a local **data base manager**. Thus we have a set of data base managers:

$$DBM = \{d_1, d_2, \ldots, d_n\}.$$

Each manager is allowed to make an update to its own copy of the data base – but then it must send a **message** to all the other managers (so that they can perform the same update on their copy of the data base). In this example we are not interested in the content of the message – but only in the header information, which describes the **sender** and the **receiver**. Thus we have the following set of messages:

$$MES = \{(s,r) \mid s,r \in DBM \wedge s \neq r\}$$

where the sender s and the receiver r are two different data base managers. When a data base manager s makes an update, it must send a message to all other managers, i.e., the following messages:

$$Mes(s) = \sum_{r \in DBM-\{s\}} 1`(s,r)$$

where the summation indicates that we form a multi-set, with n–1 elements, by adding the multi-sets $\{1`(s,r) \mid r \in DBM-\{s\}\}$, each of which contains a single element. The resulting multi-set contains one appearance of each message which has s as sender.

The equation above defines a function, Mes, which maps each data base manager to a multi-set of messages, i.e., $Mes \in [DBM \rightarrow MES_{MS}]$ (where we use A_{MS} to denote all multi-sets over A, and use $[B \rightarrow C]$ to denote all functions from B to C). Together these definitions give us the following declarations. Later in this section we shall explain why they are equivalent to the CPN ML declarations in Fig. 1.13:

constants: n : integer (* $n \geq 3$ *);

colour sets: $DBM = \{d_1, d_2, \ldots, d_n\}$;
$MES = \{(s,r) \mid s,r \in DBM \wedge s \neq r\}$;
$E = \{e\}$;

functions: $Mes(s) = \sum_{r \in DBM-\{s\}} 1`(s,r)$;

variables: s,r : DBM;

Now let us look at the net structure and the net inscriptions in Fig. 1.13. Each data base manager has three different states: *Inactive*, *Waiting* (for acknowledgments) and *Performing* (an update requested by another manager). Each message can be in four different states: *Unused*, *Sent*, *Received* and *Acknowledged*. Finally, the system can be either *Active* or *Passive*.

Initially all managers are *Inactive* and all messages are *Unused*. This is indicated by the initialization expressions, where we use the notation A to denote the multi-set which contains exactly one appearance of each element in the set A (i.e., the set A considered as a multi-set). When a manager, s, decides to make an *Update and Send Messages*, its state changes from *Inactive* to *Waiting*, while the state of its messages Mes(s) changes from *Unused* to *Sent*. Now the manager has to wait until all other managers have acknowledged the update. When one of these other managers, r, *Receives a Message*, its state changes from *Inactive* to *Performing* (the update), while the state of the corresponding message (s,r) changes from *Sent* to *Received*. Next the data base manager r may *Send an Acknowledgment* (saying that it has finished the update) and its state changes from *Performing* back to *Inactive*, while the state of the message (s,r) changes from *Received* to *Acknowledged*. When all the messages Mes(s), which were sent by the manager s, have been *Acknowledged*, the manager s may *Receive all Acknowledgments* and its state changes from *Waiting* back to *Inactive*, while the state of its messages Mes(s) changes from *Acknowledged* back to *Unused*.

To ensure consistency between the different copies of the data base this simple synchronization scheme only allows one update at a time. In other words, when a manager has initiated an update, this has to be performed by all the managers before another update can be initiated. This mutual exclusion is guaranteed by the place *Passive* (which is marked when the system is passive). Initially *Passive* contains a single e-token (as for the resource allocation system, we use e to denote "uncoloured tokens", i.e., tokens with no information attached).

It should be remarked that our description of the data base system is very high-level (and unrealistic) – in the sense that the mutual exclusion is described by means of a global mechanism (the place *Passive*). To implement the data base system on distributed hardware, the mutual exclusion must be handled by means of a "distributed mechanism". Such an implementation could be described by a more detailed CP-net – which also could model how to handle "loss of messages" and "disabled sites".

It should also be noted that the CP-net description of the data base system has several redundant places – which could be omitted without changing the behaviour (i.e., the possible sequences of steps). As an example, we can omit *Unused*. Then there will only be an explicit representation of those messages which currently are in use (i.e., in one of the states *Sent, Received* or *Acknowledged)*. We can also omit *Active* and *Performing*. It is very common to have redundant places in a CP-net, and this often makes the description easier to understand – because it gives a more detailed and more comprehensive description of the different states of the system.

Next let us use the data base system to illustrate that it is possible to move the specification of enabling constraints between a guard and the surrounding arc

```
val n = 4;
color DBM = index d with 1..n declare ms;
color PR = product DBM * DBM declare mult;
fun diff(x,y) = (x<>y);
color MES = subset PR by diff declare ms;
color E = with e;
fun Mes(s) = mult'PR(1`s,DBM-1`s);
var s, r : DBM;
```

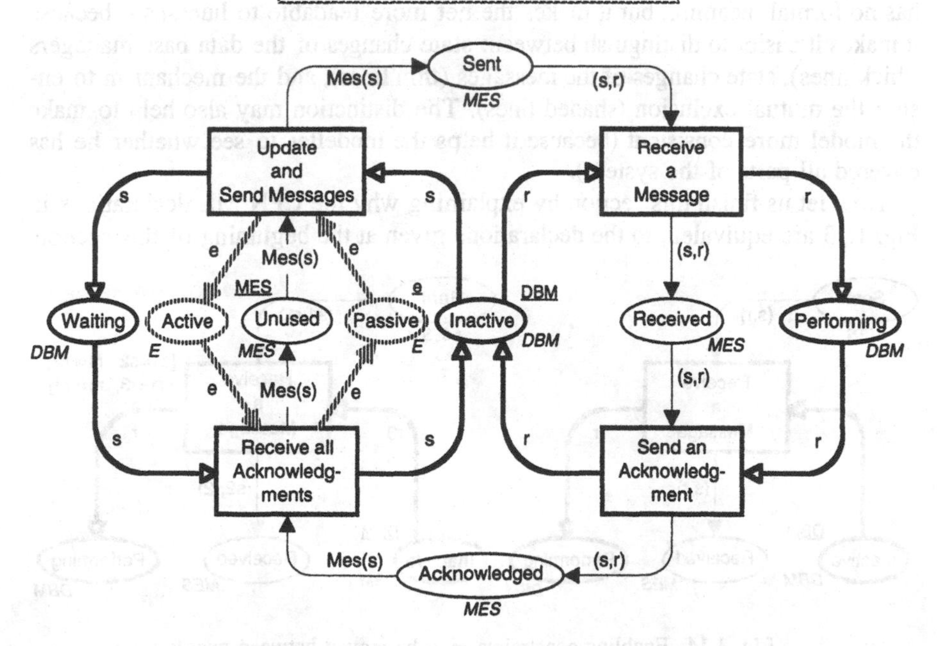

Fig. 1.13. CP-net describing the distributed data base

expressions. This can be seen in Fig. 1.14 which shows two different versions of *Receive a Message*. For the left transition we only use two variables, which appear in several arc expressions. For the right transition we use many different variables, which are related by a guard. It should be obvious that the two transitions are equivalent. A step is enabled for one of the transitions iff it is enabled for the other. Moreover, the effect of an enabled step is the same.

We shall also use the data base system to illustrate three of the very basic concepts of net theory: concurrency, conflict and causal dependency. In the initial marking of the data base system (Fig. 1.13), the transition *Update and Send Messages* is enabled for all managers, but it can only occur for one manager at a time (due to the single e-token on *Passive*). This situation is called a **conflict** (because the binding elements are individually enabled, but not concurrently enabled) – and we say that the transition *Update and Send Messages* is in **conflict with itself**.

When the transition *Update and Send Messages* has occurred for some manager s, the transition *Receive a Message* is concurrently enabled for all managers different from s. This situation is called **concurrency** – and we say that the transition *Receive a Message* is **concurrent to itself**.

The transition *Receive all Acknowledgments* is only enabled when the transition *Update and Send Messages* has occurred for some manager s, and the transitions *Receive a Message* and *Send an Acknowledgment* have occurred for all managers different from s. This situation is called **causal dependency** (because we have a binding element which can only be enabled after the occurrence of certain other binding elements).

In Fig. 1.13 we have drawn the places and arcs in three different ways. This has no formal meaning, but it makes the net more readable to humans – because it makes it easier to distinguish between: state changes of the data base managers (thick lines), state changes of the messages (thin lines), and the mechanism to ensure the mutual exclusion (shaded lines). The distinction may also help to make the model more consistent (because it helps the modeller to see whether he has covered all parts of the system).

Now let us finish this section by explaining why the CPN ML declarations in Fig. 1.13 are equivalent to the declarations given at the beginning of this section.

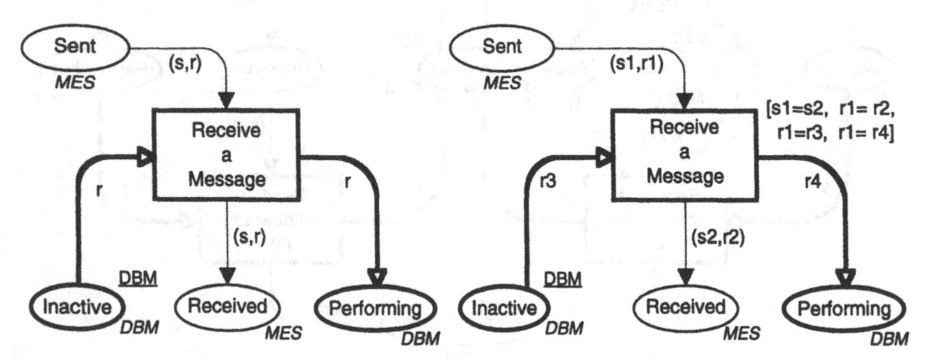

Fig. 1.14. Enabling constraints may be moved between guards and arc expressions

This explanation may be skipped by readers who are not interested in CPN ML and the CPN simulator. CPN ML is developed in the United States, and thus it uses American spelling, e.g., "color".

Line 1 declares n to be a constant, and gives it the value 4. Constants in CP-nets are used in a way which is similar to that of programming languages. This means that we can change the number of data base managers by changing the declaration of n (instead of having to change a large number of declarations and net inscriptions).

Line 2 declares the colour set DBM. This is done by means of a built-in colour set constructor which makes it is easy to declare **indexed colour sets**, i.e., sets on the form $\{x_i, x_{i+1}, \ldots, x_{k-1}, x_k\}$ where x is an identifier (i.e., a text string), while i and k are two integers (specified by means of two integer expressions). CPN ML does not recognize different font styles and thus x_r is written as x(r) or as x r (where the space after x is significant). By adding "declare ms" (where ms stands for multi-set) we instruct the CPN ML compiler to declare a constant multi-set, having a single appearance of each element in DBM. This constant is predefined, but to save space on the ML heap it is only declared when the user asks for it. The constant is denoted by ms'DBM or simply by DBM, and it is used in the initialization expression of *Inactive*. We could also have declared DBM by means of an enumeration colour set, but then it would have been impossible to make the declaration independent of the actual value of n.

Lines 3–5 declare the colour set MES. First we declare a colour set PR which is the cartesian product of DBM with itself. By adding "declare mult" we instruct the CPN ML compiler to declare a function mult'PR by which we can multiply two DBM multi-sets with each other in order to get a PR multi-set. This function is predefined, but to save space on the ML heap it is only declared when the user asks for it. As an example, we have mult'PR(2\`$d_3$+1\`d_4, 1\`$d_2$+3\`d_3) = 2\`($d_3$,$d_2$)+ 6\`(d_3,d_3) + 1\`($d_4$,$d_2$) + 3\`(d_4,d_3). The function mult'PR is used in line 7 of the declarations (see below). Next we declare a function diff$\in$[DBM$\times$DBM $\rightarrow$bool], where bool = {false, true}. The function returns true when the two arguments differ from each other, and it returns false when they are identical. Finally we declare MES to be the subset of PR which contains exactly those elements for which diff(s,r) is true. By adding "declare ms" we get a constant multi-set, denoted by ms'MES or simply by MES – and this is used in the initialization expression of *Unused*.

Line 6 declares the colour set E. This is done in exactly the same way as in Sect. 1.2.

Line 7 declares the function Mes$\in$[DBM$\rightarrow$$\text{MES}_{\text{MS}}$]. This is done by means of the predefined multiplication function mult'PR declared in line 3. The minus sign denotes multi-set subtraction.

Finally, line 8 declares the two variables s and r of type DBM.

1.4 Net Inscriptions in CPN ML

This section describes how we construct and use **net inscriptions** – in particular the arc expressions, guards, initialization expressions and colour sets. For readers who are familiar with typed lambda-calculus and functional programming languages it may be a help to know that we use types, functions, operations, variables, constants, and expressions in a way which is similar to those languages (except that for historical reasons we often use the term colour set, instead of type). In particular this means that expressions do *not* have side-effects and variables are *bound* to values (instead of being assigned to them). It also means that complex expressions are built, from variables, constants and simpler subexpressions, by means of functions and operations.

Each CP-net has a set of declarations, which by convention we position in a box with dashed lines. The declarations introduce a number of colour sets, functions, operations, variables and constants, which can be used in the net inscriptions of the corresponding CP-net, in particular in the guards, arc expressions and initialization expressions. As mentioned in Sect. 1.2, the declarations of a CP-net can be made in many different languages, e.g., by means of standard mathematical notation or by means of many-sorted sigma algebras. However, in this book we shall use a language called **CPN ML**, which is relatively close to the style of declarations used in ordinary high-level programming languages. Below we describe colour sets, functions, operations, variables and constants in more detail, and we sketch the CPN ML syntax for each of them. References to much more detailed introductions to **Standard ML** (the language upon which CPN ML is built) can be found in the bibliographical remarks.

Colour sets

Each colour set declaration introduces a new colour set, whose elements are called colours. A declared colour set can be used:

- in the declaration of other colour sets (e.g., products and records),
- in the declaration of variables (having the colour set as type),
- in the declaration of functions, operations and constants (as an example a function may map from one colour set into another colour set),
- in the colour set inscription of a place (indicating that all tokens on the place must have token colours which belong to the colour set).

Each colour set declaration implicitly declares a set of constants (the colours of the colour set). Moreover, it is often the case that the colour set declaration implicitly declares some standard functions and operations which can be used on the colours of the colour set. As an example, a colour set containing integers has the usual addition and subtraction operations.

In CPN ML colour sets can be declared in many different ways. The first possibility is to define the colour set using one of the five basic types in Standard ML (SML):

color AA = int	all integers
color BB = real	all reals

color CC = string	all text strings
color DD = bool	two colours: false and true
color EE = unit	only one colour, denoted by ().

The second possibility is to declare the colour set as being a *subset* of int, real or string:

color FF = int with 10..40	all integers between 10 and 40
color GG = real with 2.0..4.5	all reals between 2.0 and 4.5
color HH = string with "a".."z" and 3..9	all text strings with characters between a and z and length between 3 and 9.

The third possibility is to declare the colour set to be identical to bool or unit – but with other names for the colours:

color II = bool with (no,yes)	as DD, but with two different names for the colours
color JJ = unit with e	as EE, but with a different name for the colour.

The fourth possibility is to declare the colour set by explicitly specifying its colours. This can be done in two different ways:

color KK = with man | woman | child	three colours: man, woman and child
color LL = index car with 3..8	six colours: car(3), car(4), ..., car(8).

The fifth and final possibility is to declare the colour set from some already declared colour sets by means of a built-in colour set constructor. Some motivations and explanations are given immediately below the declarations, and it is recommended that these are read in parallel with the declarations. A similar remark applies to several of the other ML examples in this section.

color MM = product AA * BB * CC	all triples (a,b,c) where a∈AA, b∈BB and c∈CC
color NN = record i:AA * r:BB * s:CC	all labelled records {i=a,r=b,s=c} where a∈AA, b∈BB and c∈CC
color OO = union i1:AA + i2:AA + r:BB + c1 + c2	all colours of the form i1(a), i2(a), r(b), c1 and c2 where a∈AA and b∈BB
color PP = list AA	all lists of integers, e.g., the colour [23, 14, 3, 48]
color QQ = list AA with 3..8	as PP, but the lists must have a length which is between 3 and 8
color RR = subset AA with [2,4,6,8,10]	five colours: 2, 4, 6, 8 and 10
color SS = subset AA by Even	all even integers, i.e., all integers x for which Even(x) is true
color TT = AA	contains exactly the same colours as AA.

The difference between a **product** (MM) and a **record** (NN) is the way in which we denote the colours. In a product we use positional representation, e.g., (4, 2.7, "Petri") – and it would be illegal to write ("Petri", 4, 2.7), because from the declaration of MM we know that the first element is an integer, the second a real and the third a text string. In a record we use selectors, e.g., {i=4, r=2.7, s="Petri"}, and here it would also be legal to write {s="Petri",i=4,r=2.7}.

A **union** (OO) has a number of selectors, i1, i2, r, c1 and c2. Some of the selectors, i1, i2 and r, have an attached colour set, and each of them contributes to the union colour set with as many colours as the corresponding selector colour set has. Such a colour is denoted by "sel(col)" or "sel col", where sel is the selector and col the colour of the selector colour set. Other selectors, c1 and c2, have no attached colour set, and each of them contributes to the union colour set with a single colour, which is denoted by the selector itself. Notice that both "i1(4)" and "i2(4)" are legal colours. They are both constructed from 4, but by means of two different selectors, i1 and i2. A union constructed in this way is called a *disjoint union*.

A **subset** (RR and SS) can be defined in two different ways. The first possibility is to list the colours which should be included (these colours must of course belong to AA). The second possibility is to use a predicate, i.e., a function mapping from AA into bool = {false, true}. In the latter case the subset will contain exactly those colours in AA which are mapped into true.

The product, record, union, list and subset constructors can be arbitrarily nested. As an example, we may declare the following colour sets (which we hope are self-explanatory):

```
color Name = string;
color NameList = list Name;
color Year = int;
color Month = with Jan | Feb | Mar | Apr | May | Jun | Jul | Aug | Sep | Oct | Nov | Dec;
color Day = int with 1..31;
color Date = product Year * Month * Day;
color Person = record name : Name * birthday : Date * children : NameList;
```

Functions

Each function declaration introduces a function. The function is not allowed to have side effects (this means that it is evaluated without influencing any other part of the system). The function takes a number of arguments and returns a result. The arguments and the result have a type which is either a declared colour set, the set of all multi-sets over a declared colour set, or some other type recognized by CPN ML, e.g., bool. A declared function can be used:

- in the declaration of colour sets (e.g., to construct subset colour sets),
- in the declaration of other functions, operations and constants,
- in arc expressions, guards and initialization expressions.

In CPN ML functions are declared as indicated by the following examples:

```
fun Fac(n:AA) = if n>1 then n*Fac(n–1) else 1
fun Even(n:AA) = ((n mod 2) = 0)
fun F1(x:U) = case x of p => 2`e | q => 1`e
fun Div(n:AA, 0:AA) = 0 | Div(n,m) = n div m
fun Rev([]:PP) = [] | Rev(head::tail) = Rev(tail)^^[head].
```

Fac calculates the factorial of n. Even tells whether the argument is even or not (it was used in the declaration of SS, which means that the declaration of Even must precede the declaration of SS). F1 calculates a multi-set of E-tokens. It was used in Fig. 1.10 (and the colour sets U and E are as declared there). Div is a little more complicated, because it is defined using two clauses, separated by "|". If the two arguments match the pattern in the first clause (which means that the second argument is 0) Div returns 0. Otherwise, the arguments will match the pattern in the second clause, and Div returns the value of "n div m" (where div is the standard integer division operation). Notice that the order of the two function clauses is significant. If they are interchanged the call Div(3,0) will fail (i.e., raise an exception) because "3 div 0" is undefined. Finally, Rev is a recursive function. It takes a list of integers as argument and it returns a result which is the reversed list. If the argument is the empty list [] the result is also empty. Otherwise, the argument matches the expression head::tail – where head is the first element of the list while tail is the rest of the list – and a recursive call of Rev plus the list concatenation operator ^^ finishes the job. It should be remarked that the Rev function is rather inefficient because its execution time grows as the square of the length of the argument. It is also possible in CPN ML (of course) to make a more efficient version of Rev, where the execution time grows linearly.

Above we have only specified the types of the function arguments. The types of the function results will then be automatically calculated by the CPN ML compiler. However, it is also possible to specify the result type, and it will then be checked that the result type is consistent with the declaration:

```
fun Fac(n:AA):AA = if n>1 then n*Fac(n-1) else 1
fun Even(n:AA):bool = ((n mod 2) = 0)
fun F1(x:U):E ms = case x of p=> 2`e | q => 1`e
fun Div(n:AA, 0:AA):AA = 0 | Div(n,m) = n div m
fun Rev([ ]:PP):PP = [ ] | Rev(head::tail) = Rev(tail)^^[head].
```

We have now specified that the result of Fac is of type AA, while the result of Even is of type bool. Analogously, we have specified that the result of F1 is of type "E ms", which denotes the type containing all multi-sets over E. This type is constructed from the colour set E by using the built-in type constructor "ms". The ms-constructor can be applied to all colour sets (and all other types recognized by CPN ML, as long as these have an equality operation). It is not possible to use the ms-constructor to declare new colour sets. Finally, we have specified that the result of Div is of type AA, while the result of Rev is of type PP.

It is also possible to go in the other direction, and leave out both the types of the arguments and the type of the result. Then the CPN ML compiler will try to determine the missing types itself. However, this may not always be possible (and the CPN ML compiler will then issue an error message):

```
fun Fac(n) = if n>1 then n*Fac(n-1) else 1
fun Even(n) = ((n mod 2) = 0)
fun F1(x) = case x of p=> 2`e | q => 1`e
```

```
fun Div(n,0) = 0 | Div(n,m) = n div m
fun Rev([ ]) = [ ] | Rev(head::tail) = Rev(tail)^^[head].
```

For Fac the CPN ML compiler will first conclude that the argument n must be an integer, because of the > operation used on 1 (notice that 1 is an integer, while 1.0 would be a real). Next the compiler will conclude that the result of Fac is of type int (because it is either the integer 1 or the result of multiplying the integer n with something else (which then also must be an integer)). For Even the CPN ML compiler will conclude that the argument n must be an integer (because of the modulo operation). Then the compiler will conclude that the result must be a boolean (because of the = operation). For F1 the compiler will conclude that the argument x is of type U (because x in the case-expression is compared with p and q which are both of type U). Next the compiler will conclude that the result is of type E ms (because of the ` operation used on e). For Div the compiler will conclude that both the arguments and the result are of type int. For Rev the situation is more interesting. There is no way the compiler can find out what kind of lists the arguments are of. Thus the compiler concludes that the function can be used for *all kinds* of lists, and this means that the function becomes **polymorphic** (i.e., can take arguments of different types).

Being a functional language SML has a lot of different facilities for the declaration of functions – and all of these can be used in CPN ML. In the above, we have shown some of the possibilities, and we shall introduce some of the others as they become necessary for our examples.

Operations

The only difference between functions and operations is that the former uses prefix notation, such as "Div(14,3)", while the latter uses infix notation, such as "14 Div 3". In CPN ML a declared function can be turned into an operation by an infix directive:

infix Div	changes Div from a function to an operation.

It is then legal to write "14 Div 3", and it is illegal to write "Div(14,3)". It is possible to specify the precedence of the new operator, and tell whether it associates to the left or to the right (for details see a description of SML).

Variables

Each variable declaration introduces one or more variables, with a type which must be an already declared colour set. A declared variable can be used:

- in arc expressions and guards (but *not* in initialization expressions).

In CPN ML variables are declared as indicated by the following examples:

var no : AA	declares a variable of type AA
var x,y,z : AA	declares three variables of type AA.

Constants

Each constant declaration introduces a constant, with a type which must be an already declared colour set – or some other type recognized by CPN ML, such as int. A declared constant can be used:

- in the declaration of colour sets (e.g., to construct indexed colour sets),
- in the declaration of functions, operations and other constants,
- in arc expressions, guards and initialization expressions.

In CPN ML constants are declared as indicated by the following examples:

val car4 = car(4)	declares a constant of type LL
val car5 = car(5) : LL	declares a constant of type LL
val n = 4	declares a constant of type int.

In the first and third declarations the type of the constant is calculated by the CPN ML compiler. In the second it is specified by the modeller and checked by the compiler.

Net expressions

Each net expression is built from a number of variables, constants, operations and functions. Net expressions are used:

- as arc expressions, guards and initialization expressions.

Each net expression can be **evaluated** with respect to a **binding** of its variables. A binding associates with each variable in the expression a constant of the same type as the variable. To perform the evaluation we first modify the expression by replacing each appearance of a variable with the corresponding constant. Then we apply the operations and functions and this yields a constant which is called the **value** of the expression (with respect to the corresponding binding). Each expression has a **type**, which contains all possible evaluation values.

The type of a guard is bool = {false,true}, while the type of an arc/ initialization expression is C ms where C is the colour set of the corresponding place. As a shorthand, we also allow the type of an arc/initialization expression Expr to be C, and we then replace Expr by 1`(Expr), which is of type C ms. In CPN ML expressions are constructed as indicated by the following examples:

x=q	x is a variable; q is a constant; the type is bool; <x=p> evaluates to false while <x=q> evaluates to true
(x,i)	x and i are variables; the tuple-construction is an operation; the type is P (cf. Fig. 1.7); <x=p,i=0> evaluates to (p,0) while <x=q,i=37> evaluates to (q,37)
1`(x,i)	x and i are variables; 1 is a constant; ` and the tuple-construction are operations; the type is P ms; <x=p,i=0> evaluates to 1`(p,0) while <x=q,i=37> evaluates to 1`(q,37)
3`(q,0)	3, q and 0 are constants; ` and the tuple-construction are operations; the type is P ms; the empty binding <> evaluates to 3`(q,0) and so do all other possible bindings

2`e	2 and e are constants; ` is an operation; the type is E ms; <> evaluates to 2`e and so do all other possible bindings
case x of p => 2`e | q => 1`e	x is a variable; p, q, e, 1, and 2 are constants; ` and the case-construction are operations; the type is E ms; <x=p,i=0> evaluates to 2`e while <x=q,i=37> evaluates to 1`e
if x=q then 1`(q,i+1) else empty	x and i are variables; q, 1 and empty are constants; `, =, +, the tuple-construction and the if-construction are operations; the type is P ms; <x=p,i=0> evaluates to empty while <x=q,i=37> evaluates to 1`(q,38)
Mes(s)	s is a variable; Mes is a function; the type is MES ms (cf. Fig. 1.13); <s=d_1> evaluates to 1`($d_1$,$d_2$) + 1`(d_1,d_3) + ... + 1`($d_1$,$d_{n-1}$) + 1`(d_1,d_n)
let n=5 in n * x + 2 end	x is a variable; n is bound by the let-construction and thus it is *not* a variable of the expression; 2 and 5 are constants; +, * and = are operations; the type is int; <x=3> evaluates to 17.

SML has many other facilities for the construction of expressions. Later, we shall introduce some of these, as they become necessary for our examples.

1.5 Construction of CPN Models

As for all other modelling languages, it takes a considerable amount of experience to become a good and efficient CPN modeller. The best way to get started is to play with small and medium-size models created by others. Try to understand the model, and try to make simple modifications of it. If possible choose a model from an application area which you are familiar with or – even better – the application area for which you plan to use CP-nets. This will usually mean that you can reuse some of the constructions when you start on your own models. This section contains a number of hints and suggestions, which may make it easier for the inexperienced modeller to get started. Some of the guidelines are general – in the sense that they apply also to the creation of many other kinds of structured descriptions, e.g., ordinary programs and the kinds of diagrams used in structured analysis. Other guidelines are more particular for CP-nets.

1. Start by identifying some of the most important components of the modelled system. This can be done by creating a number of lists, e.g., containing objects, processes, states and actions of the system. From the beginning it may not always be clear to what category each item belongs. Thus the modeller should continue to revise the lists and try to make the categorisation as consistent as possible. He should also try to make the lists as complete as possible. For this purpose it is a good idea to glance through all the available descriptions of the system (informal or formal), and from these try to extract components of the four categories mentioned above. As an example, Fig. 1.15 shows how the initial lists might look when we try to model a telephone system (cf. Exercise 1.8).

2. Consider the purpose of your model and determine an adequate level of detail. Many actions consist of a number of subactions, and it must be decided to what extent it is relevant to model these as separate actions. As an example, a primitive action of the telephone system may be the dialling of a phone number (as indicated in the action list of Fig. 1.15). However, we may want to investigate, in detail, the functioning of the phone buttons – or the component of the phone translating the pressing of the buttons into a sequence of electrical signals (encoding the dialled phone number). If this is the case, we probably want to decompose the dialling action into a number of subactions, such as the pressing of a button for each individual digit (or even the pressing and subsequent release of the button). It is only when we consider the purpose of the modelling activity that it becomes possible to determine that one of these levels of details is more adequate than the others.

3. Try to find good mnemonic names for objects, processes, states and actions. Add these names to the lists. The names should make it easy to remember what the individual elements represent – and they should be so short that they can be used in the CP-net, e.g., as place and transition names. Investing some time to find adequate names will make the development of the CP-net easier, because one can immediately remember what the individual net elements represent. Moreover, the use of good mnemonic names makes it much easier for other people to read and understand the CP-net. One should avoid using the same name too often, or providing the same information twice, by having it both in the place and transition names. As an example, it is of little help if many places are called *Waiting*, or *Waiting for XX* where *XX* is the name of the subsequent transition.

4. Do not attempt to cover all aspects of the considered system in the first version of your model. CP-nets can be used to create very complex models – as indicated by the industrial applications described in Chap. 7. To get started,

Objects

- Phones (with a receiver, a set of buttons and a buzzer).
- Human beings (using the phones).
- Telephone network (with wires, computers, etc.).

Processes

- Senders (i.e., phones from which another phone is being called).
- Recipients (i.e., phones which are being called from another phone).
- Telephone exchange (e.g., checking whether a requested connection can be established).

Actions (at a phone)

- Lift the receiver.
- Dial a number.
- Put down the receiver.
- Talk

States (of a phone)

- Inactive.
- Continuous tone (indicating that a number may be dialled).
- Tone with short intervals (indicating that the recipient is engaged).
- Ringing (indicating that someone is calling the phone).

States (of a connection)

- Request (a number has been dialled, but it has not been checked whether the recipient is inactive).
- Call (it has been checked that the recipient is inactive, but no one has yet lifted the receiver).
- A full connection has been established.

Fig. 1.15. Component lists for a telephone system

however, it is often necessary to make as many simplifications as possible, even though some of these may be unrealistic and unacceptable for the final model. As an example, a first version of the telephone system may omit time-outs and neglect all special services (such as conference calls and the existence of local switchboards). Analogously, for a communication protocol we may assume that no messages are lost or distorted. When the first model has been created and thoroughly tested, by simulation and/or formal analysis, we may then gradually introduce some of the omitted aspects. In this way we distribute the mental complexity of the modelling task over the different iterations. This means that it becomes easier (and faster) to obtain the first models. However, experience shows that this method also reduces the total modelling time.

5. Choose one of the processes in the modelled system and try to make an isolated net for this process. Represent each state by a place, and each possible change from one state to another by a transition. While doing this, it is usually a good idea to consider two states to be identical if they have the same logical properties – even though they may have slightly different values of data. Analogously, we usually identify actions which have the same purpose and effect – even though they may operate on slightly different states and data. As an example, the telephone system may have a place representing all those states in which a phone number has been dialled (although these states are slightly different, because different numbers were dialled). Analogously, we may have a single transition representing all the actions which bring a phone to one of these states (i.e., all the different dialling actions). The processes representing calling phones may look as indicated in Fig. 1.16 (where we have shown only the upper part of the net). A token on one of the places indicates that the call process is in one of the corresponding states. We have assumed that there is exactly one caller. This means that we ignore (for the moment) how a phone shifts between being inactive, being a sender (i.e., initiating a call) and being a recipient (i.e., receiving a call). It also means that it is less important to be precise about the mechanism by

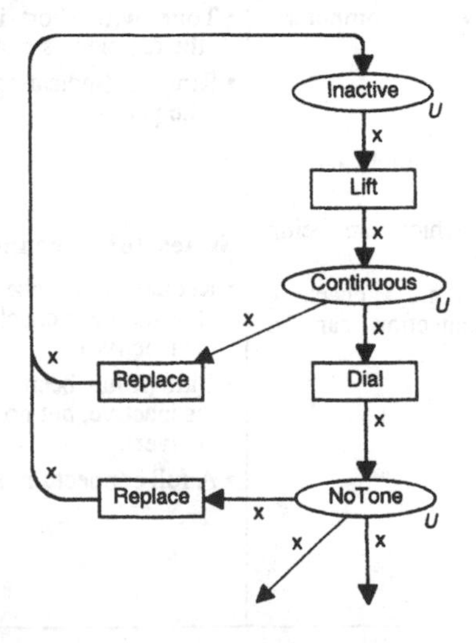

Fig. 1.16. Isolated net describing calling phones

which we distinguish between different phones. These details will be added later – when we combine the net of Fig. 1.16 with other nets representing other kinds of processes. When the modelled process is sequential the net will be a state machine. This means that each transition removes exactly one token (from the place of the predecessor state) and adds one token (to the place of the successor state).

6. Use the net structure to model control and the net inscriptions to model data manipulations. It is often a good idea to use the net structure (i.e., the places, transitions and arcs) in a similar way as the control structures of an ordinary programming language, while the net inscriptions are used for the detailed data manipulations. Figure 1.17 indicates how some of the usual control structures can be represented in a CP-net. In the first net the occurrence of transition T1 is followed by the occurrence of either T2, T3 or T4. The choice between these three alternatives may be non-deterministic or it may be governed by guards attached to T2, T3 and T4. The guards may test the colour of the token removed from A2. With a more complex net structure, the guards may also test the colour of tokens at other input places of T2, T3 and T4. The representation of the IF, WHILE and REPEAT statements should be straightforward. As before, the guards (i.e., the boolean condition b) may involve the colours of tokens at the different input places of the corresponding transitions. The procedure call and the process invocation are similar to each other. The main difference is that the procedure call postpones any further actions of the caller until the procedure has finished the execution and returned the results, while the process invocation immediately allows the initiating process to continue.

7. Distinguish between different kinds of tokens. When we interpret a net it is often possible to distinguish between different kinds of tokens used to model different kinds of phenomena. Some tokens represent the logical state of processes, while others represent moving physical objects (such as documents and the carts of a transport system). Some tokens represent messages, shared data or shared physical resources (such as machines and tools). The above is not intended to be an exhaustive list. The different kinds of tokens will vary from one application area to another. By identifying them (and by highlighting the corresponding places in different ways, as explained in Sect. 1.6), the modeller may greatly enhance the consistency and the readability of the model.

8. Use different kinds of colour sets. It is important to know how to use the different kinds of colour sets – such as indexed sets, cartesian products and lists. CPN ML contains a number of different colour set constructors. Enumeration colour sets are often used to identify processes or tasks. They are used when we want to name a relatively small number of objects by means of mnemonic names (e.g., move, stop, open_door, etc.). Indexed colour sets are used for similar purposes – but in the case where we have a class of objects which we want to denote by a class name and an object number (e.g., car(1), car(2), ..., car(8)). Product and record colour sets are used to represent sets of attributes (e.g., the sender, receiver, number and data contents of a message). The main difference between these two kinds of colour sets is the way in which we access the individual components. For product colour sets we use a positional notation, while in record colours sets we use selectors with mnemonic names. The choice between these

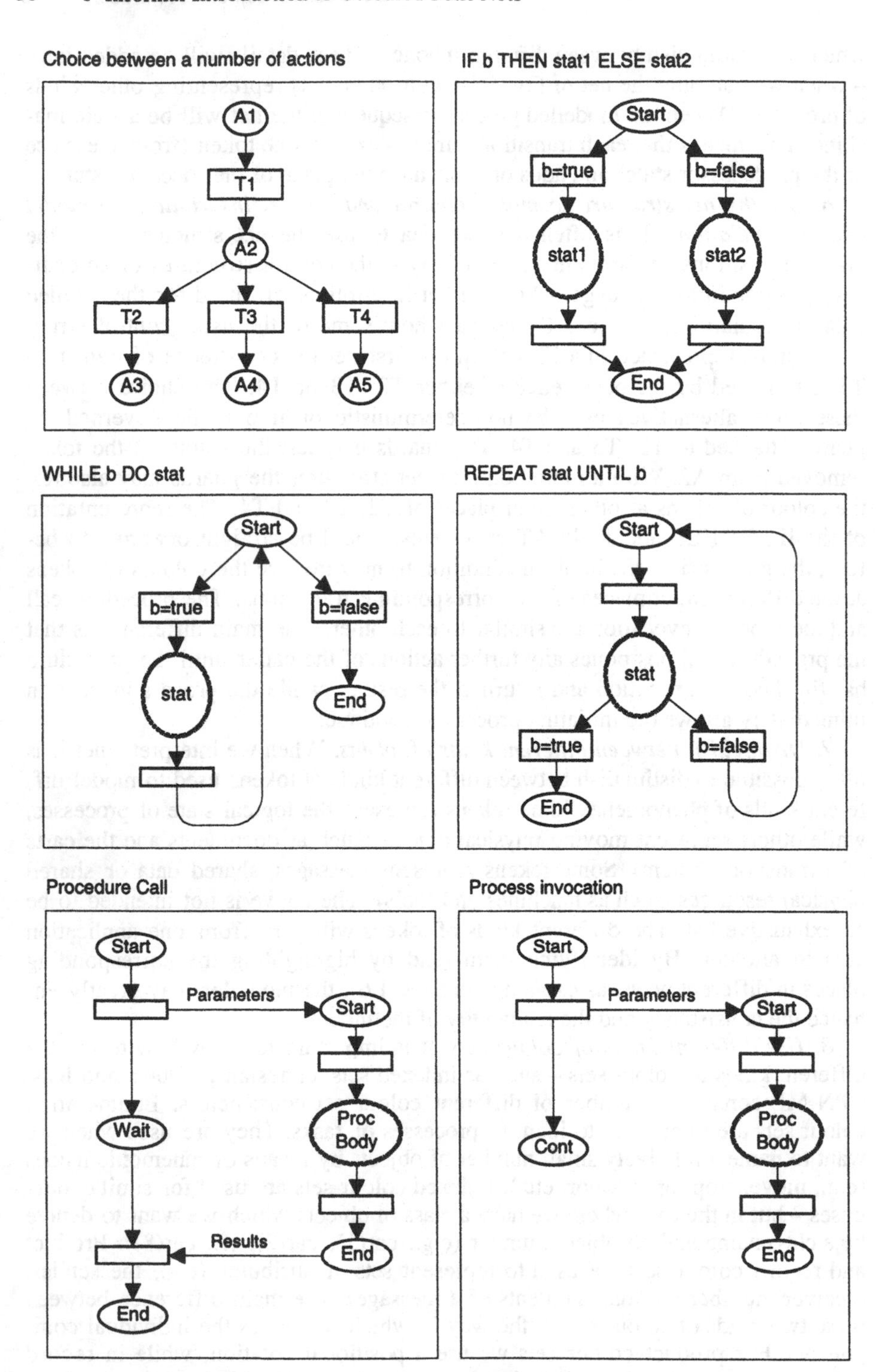

Fig. 1.17. How to represent different kinds of control structures

two colour set constructors depends to a very large degree upon the personal preference of the modeller and upon the traditions of the application area. Union colour sets are used when the same place may contain more than one kind of tokens. They are in some respects similar to the use of variant records in ordinary programming languages. Finally, list colour sets are used in a similar way as arrays, i.e., when we have a fixed or varying number of attributes which are all of the same type. There is, however, another very important use of list colour sets. When we want to implement a queue, we often do this by representing all the queued objects in a single token, which we then manipulate as a queue. This means that objects are added to the end and removed from the front of the list – as indicated by Fig. 1.18. A stack can, of course, be obtained in a similar way. If you use an inscription language which is different from CPN ML, there may of course be other colour set constructors.

9. Augment the process net by describing how the process communicates/interacts with other processes. This can be done in many different ways. Asynchronous communication is described by adding an output arc for each outgoing message and an input arc for each incoming message. Figure 1.19 illustrates how this is done (with and without acknowledgment). Synchronous communication is described by merging the transitions which represent the communicating actions. However, as shown in Fig. 1.20, it may first be necessary to duplicate some of the transitions. This happens when a single action can communicate with several alternative actions. Sharing of data is described by a shared place, which is both input and output for each action using the data. When the place has a single token, one process can use the data at a time. However, it is also possible to represent the data by a set of tokens. This means that several processes can use (different parts of) the data simultaneously. Synchronization imposed by shared resources is modelled in a similar way as synchronization by shared data – except that the resources are usually taken by one transition and returned by another (as illustrated by the resource allocation system in Fig. 1.7). As an example, Fig. 1.21 shows how the calling phones of Fig. 1.16 communicate with the telephone exchange by creating and removing requests.

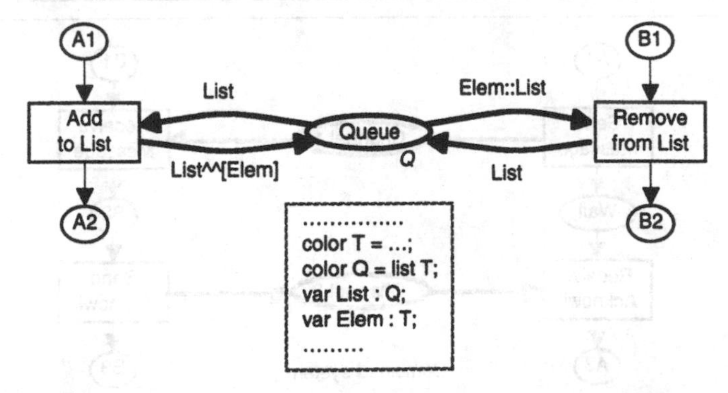

Fig. 1.18. How to represent a queue by means of a list colour set

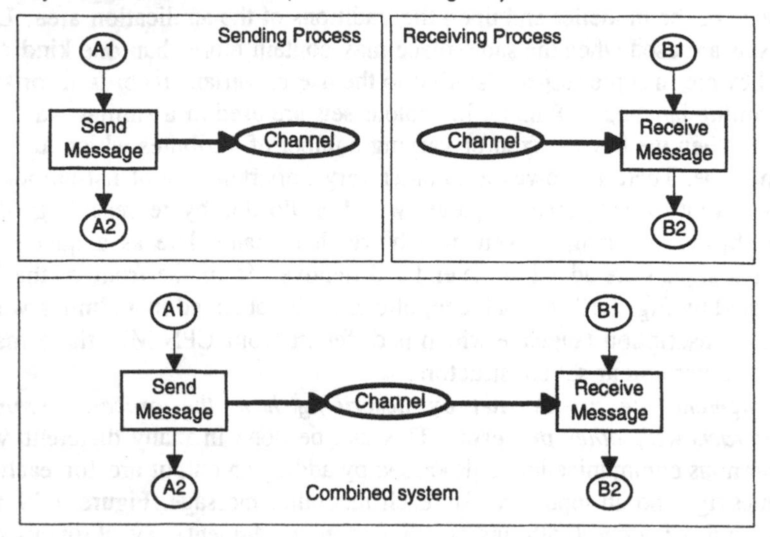

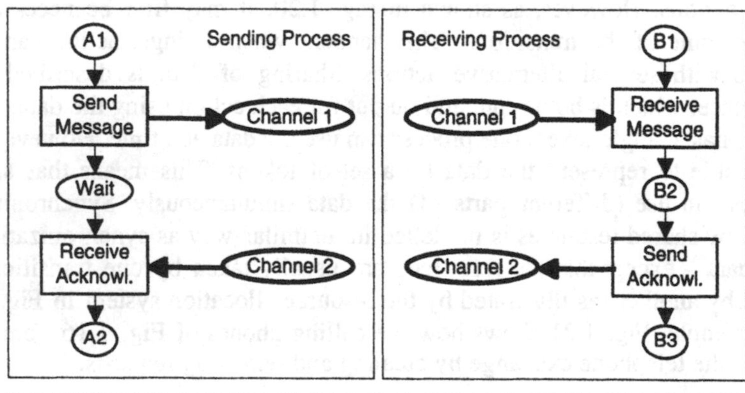

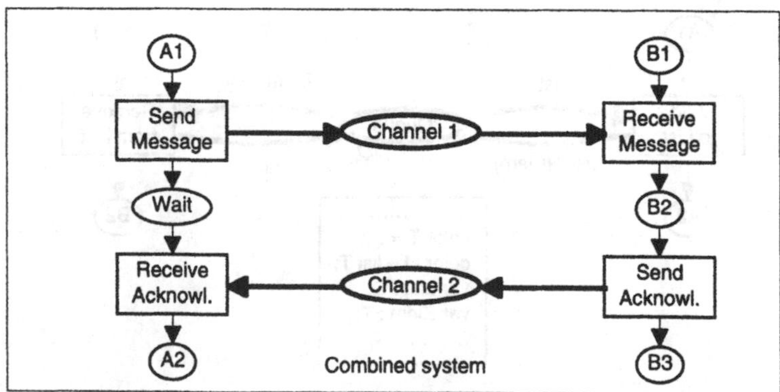

Fig. 1.19. Different kinds of interactions

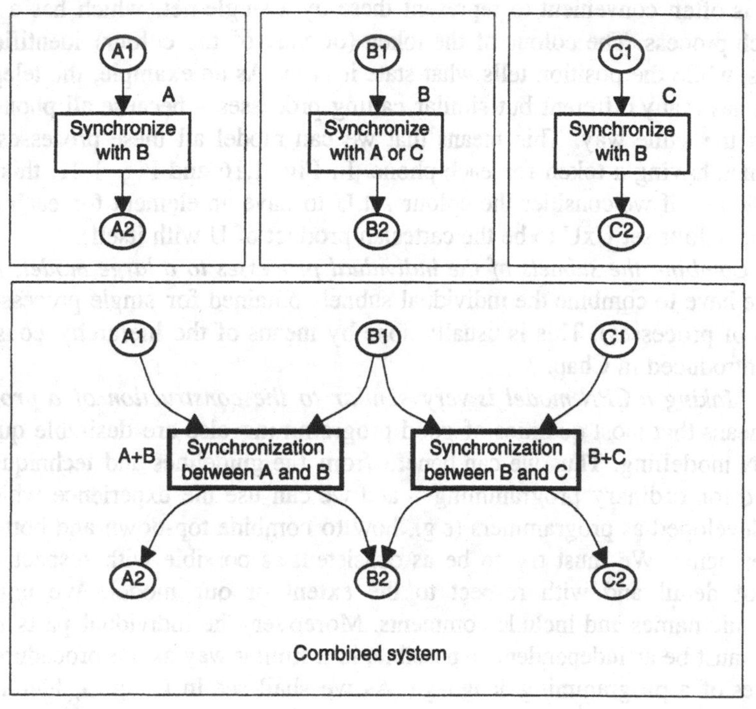

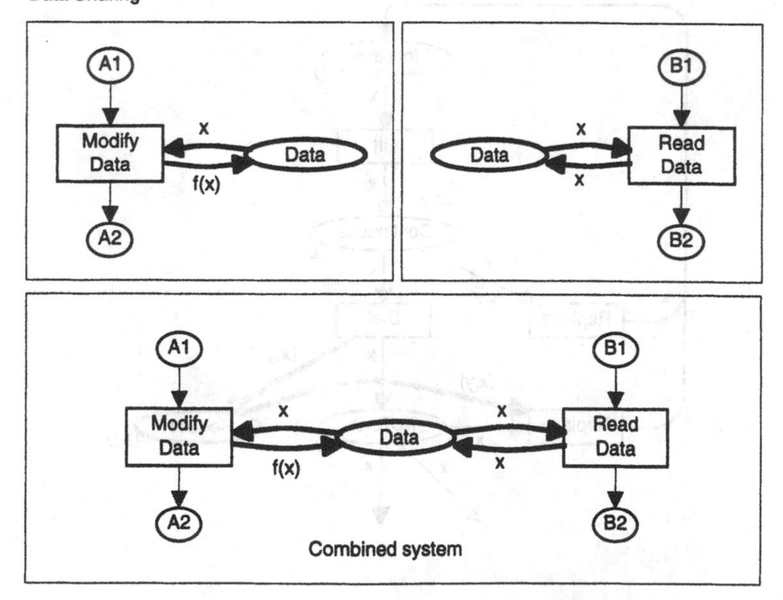

Fig. 1.20. Different kinds of interactions (continued)

10. Investigate whether there are classes of similar processes. When this is the case it is often convenient to represent these by a single net, which has a token for each process. The colour of the token (or part of the colour) identifies the process, while the position tells what state it is in. As an example, the telephone system has many different but similar calling processes – because all phones behave in the same way. This means that we can model all these processes by a single net, having a token for each phone. In Fig. 1.16 and Fig. 1.21, this is already done – if we consider the colour set U to have an element for each phone (and the colour set UxU to be the cartesian product of U with itself).

11. Combine the subnets of the individual processes to a large model. At the end, we have to combine the individual subnets obtained for single processes (or classes of processes). This is usually done by means of the hierarchy constructs to be introduced in Chap. 3.

12. Making a CPN model is very similar to the construction of a program. This means that most qualities of good programming also are desirable qualities of CPN modelling. Thus we can benefit from the guidelines and techniques developed for ordinary programming – and we can use the experience which we have developed as programmers (e.g., how to combine top-down and bottom-up development). We must try to be as consistent as possible with respect to the level of detail and with respect to the extent of our model. We must use mnemonic names and include comments. Moreover, the individual parts of our model must be as independent as possible, in a similar way as the procedures and modules of a programming language. As we shall see in Chap. 3, hierarchical CP-nets provide powerful mechanisms to obtain and combine such modules.

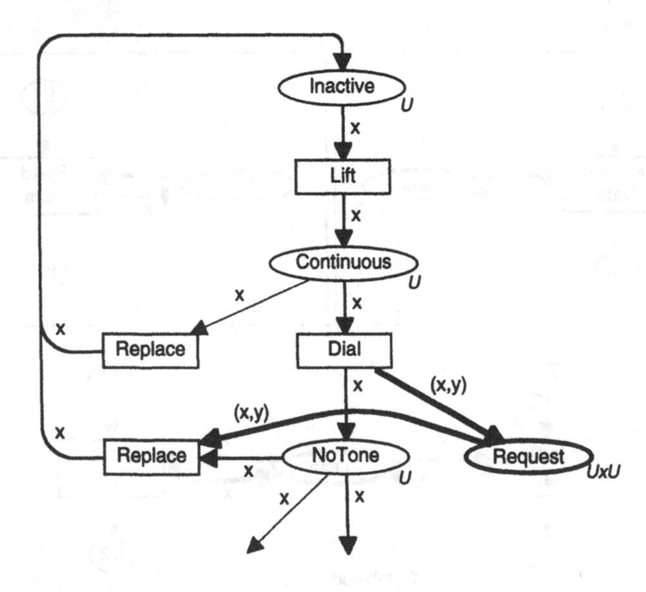

Fig. 1.21. Calling phones and their interaction with other processes

1.6 Drawing of CPN Diagrams

It takes a lot of experience, care and thought to be able to create nice and readable CP-nets, and there is no ultimate set of rules saying how CP-nets should look. The situation is analogous to programming, and we can actually learn much by considering the qualities which a readable program should possess:

- mnemonic names,
- consistent and transparent indentation strategy,
- comments,
- consistency.

The intention of this section is to give a few guidelines, which – although primitive and sometimes straightforward – may help an inexperienced modeller to create better CP-nets. The guidelines will lead to more readable CP-nets. However, it is also important that each group of modellers develops their own style and standards. This will mean that from the layout, naming and comments of a CP-net, they will be able to recognize submodels that are identical or similar to what they have dealt with before.

It is important to choose good **mnemonic names** for places, transitions, colour sets, functions, operations, variables and constants. To save space some modellers use non-mnemonic names for the places and transitions (e.g., P1, P2,... and T1, T2,...). They then add a table relating these non-mnemonic names to a short textual description of the corresponding states and actions. This saves space in the CPN diagram, but it makes the diagram very difficult to read for other people – because they constantly have to turn to the table to see what the different nodes represent. A much better idea is to use full mnemonic names (e.g., *Update and Send Messages* and *Receive all Acknowledgments*) or abbreviated mnemonic names (e.g., *UpdSendMes* and *RecAllAck*). When abbreviated mnemonic names are used, a table with a more lengthy explanation of each name should be added. This can be part of the textual explanation which accompanies the CPN diagram, but it is often a better idea to insert it directly into the CP-net (as one or more auxiliary nodes – see below).

It is important to use **textual comments** in declarations and lengthy net inscriptions. A CPN ML comment is a text string surrounded by (* and *), and it can be inserted at any position where a space can be inserted.

For CP-nets it is also possible to use **graphical comments**, i.e., to create **auxiliary** graphical objects. The auxiliary objects are part of the CPN diagram, but they do not influence the behaviour of the CP-net. Auxiliary objects should be given a position and form which makes it easy to distinguish them from the members of the net structure and from the net inscriptions. Auxiliary objects can be used for many different purposes. As described above, an auxiliary node may offer a more lengthy explanation of abbreviated place and transition name. As a second example, an auxiliary node may enclose a number of related places and transitions. As a third example, it is often difficult to position the small circles and text strings, representing the current marking, next to the corresponding places – because they then will be on top of other objects. Instead we can position them at some distance and connect each of them to the corresponding place with an auxiliary connector.

The rest of this section will deal with the layout of the net structure, i.e., how to choose the position, shape, size and shade of places, transitions, arcs and net inscriptions. We shall give ten simple drawing rules and illustrate each of them with one or more graphs showing how a small subnet looks when the rule is followed/ignored. To make our point we shall often exaggerate the errors. We shall omit the net inscriptions (when they have no importance for the corresponding rule). Notice that several of the rules are potentially conflicting, in the sense that the fulfilment of one of them may make it difficult to fulfil another. Then the modeller must decide which of the rules, in the given situation, it is most important to follow.

As a "meta-rule" it should be noted that a beautiful and aesthetically pleasing CP-net is usually also a readable net, and vice versa. Furthermore, it is recommended to use as many of the following graphical effects as possible:

- Position (relative to the other objects),
- Shape (rectangle/ellipse/rounded box, curving/straight arcs, different forms of arrowheads, etc.),
- Size (of nodes, text and arrowheads),
- Shading (of the lines, the interiors, and the arrowheads),
- Line thickness (of nodes and arcs),
- Text appearance (font type, size, style, alignment),
- Colour (of lines, interiors and texts),
- Layering of objects (so that some of them are on top of others).

We now present our ten rules:

1. Collect all inputs arcs on one side of the node and all output arcs on the opposite side. This will make it much easier to see how nodes are related to each other. In Fig. 1.22 the rule is illustrated for a transition. The same rule also applies to places, but here it is probably of less importance.

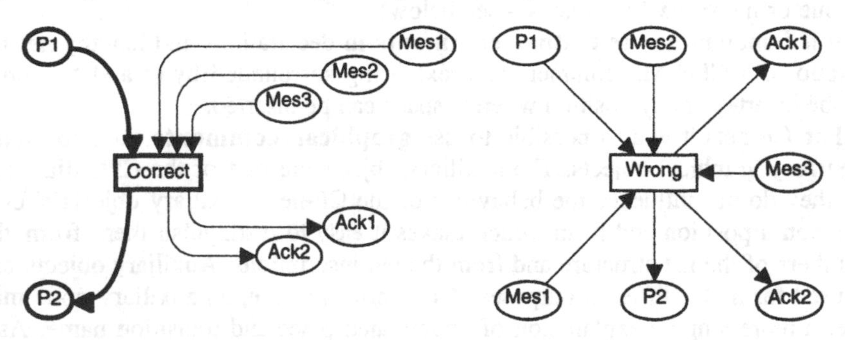

Fig. 1.22. Rule 1: Collect inputs arcs on one side of the node and output arcs on the opposite side

2. *Position related nodes and arcs on a figure with a nice geometrical shape*, such as a circle, ellipse or rounded box. Moreover, it is a good idea to **highlight** each such figure. This means that we use a special shading, line thickness and/or colour for the corresponding nodes and arcs. It should also be noted that it is much easier for the eye to follow curving arcs than arcs which have sharp bending points.

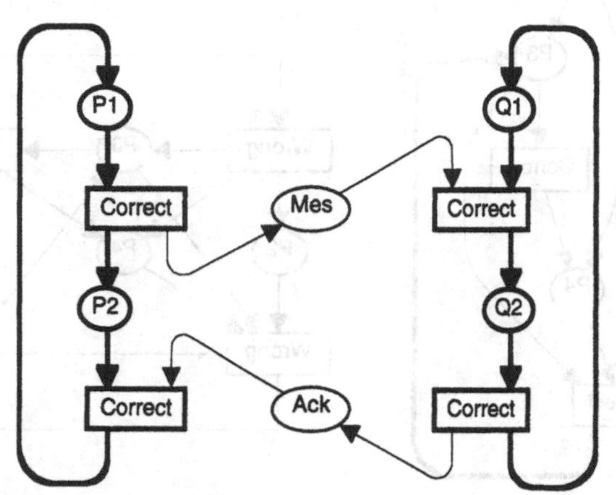

Fig. 1.23. Rule 2: Position net elements on nice figures; use highlighting and curving arcs

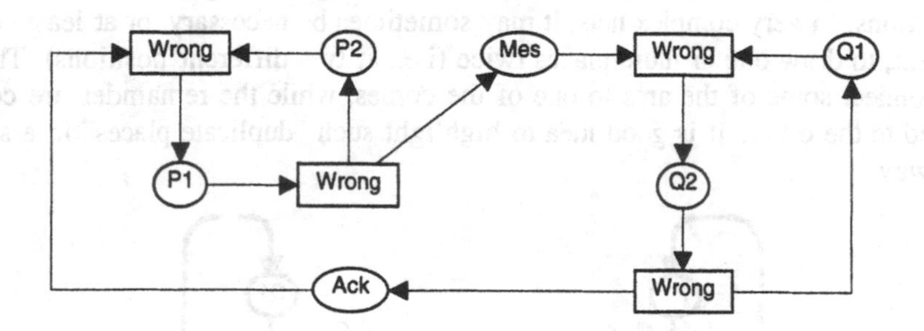

Fig. 1.24. Rule 2 (cont.): Same subnet as in Fig. 1.23

3. *Keep a main flow direction.* Usually this is from top to bottom, or from left to right. Arcs with a direction which is against the main flow may be highlighted.

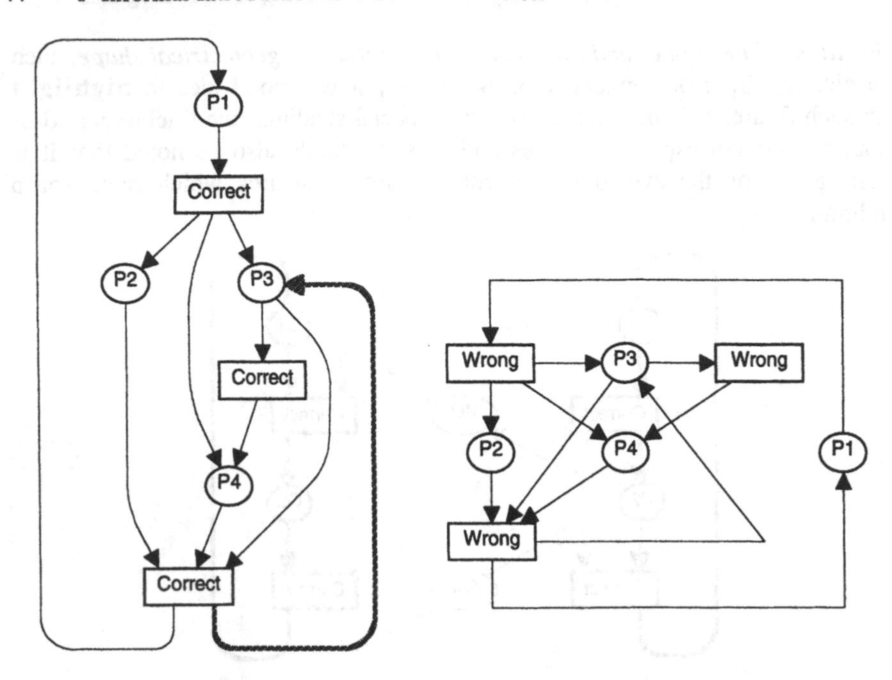

Fig. 1.25. Rule 3: Keep a main direction of flow

4. Avoid crossing arcs. This can often be done by moving a few places and transitions. In very complex nets, it may sometimes be necessary, or at least convenient, to draw one or more places twice (i.e., at two different positions). Then we connect some of the arcs to one of the copies, while the remainder are connected to the other. It is good idea to highlight such "duplicate places" in a special way.

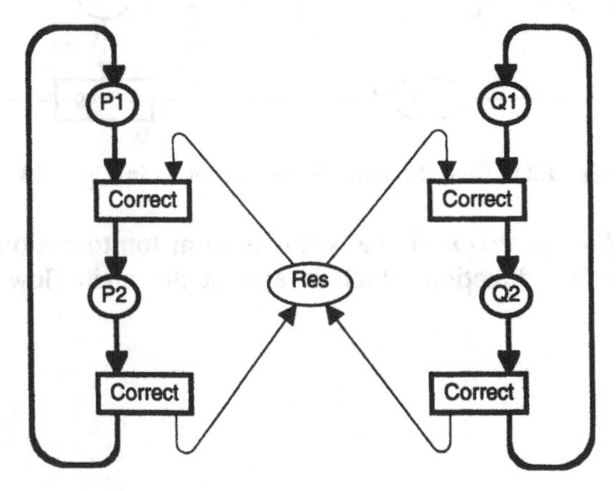

Fig. 1.26. Rule 4: Avoid crossing arcs when possible

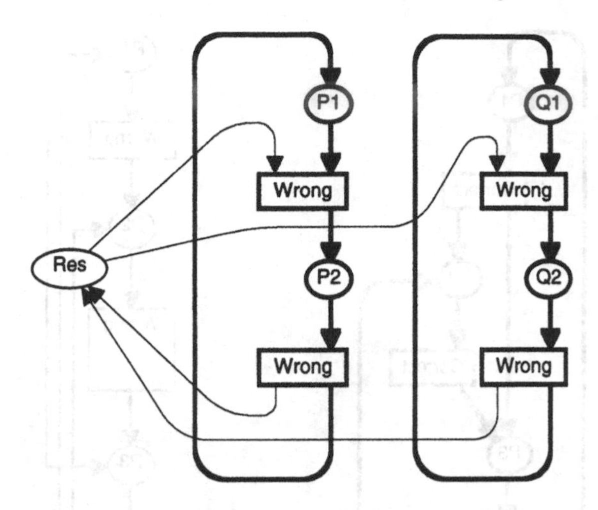

Fig. 1.27. Rule 4 (cont.): Same subnet as in Fig. 1.26

5. *Use arcs which are as simple and regular as possible.* This means that we should avoid arcs with unnecessarily complex shapes, arcs which have sharp corners, and arcs which come too close to nodes and other arcs. If two arcs must cross they should do this at a right angle (or close to it). Arc corners on top of other arcs should be avoided. A double-headed arc can be used when a pair of nodes have two connecting arcs with opposite direction and the same arc expression. This convention makes it easy to see that the transition does not change the marking of the place.

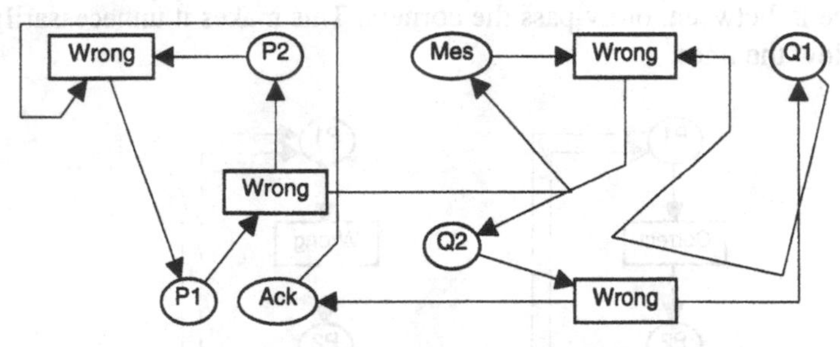

Fig. 1.28. Rule 5: Keep the arcs as simple and regular as possible (same subnet as in Figs. 1.23 and 1.24)

6. *Avoid arcs which run parallel to each other with a short distance between them.* Violations of this rule are dangerous, in particular, when the arcs have different directions. If the arcs run in the same direction, and have a common source or a common destination it is usually a good idea to position parts of the arcs on top of each other (as illustrated by the two rightmost arcs in Fig. 1.7). If the arc expressions are identical, a single copy can be positioned at the common segment of the arcs. Otherwise the arc expressions must be positioned at the private segments.

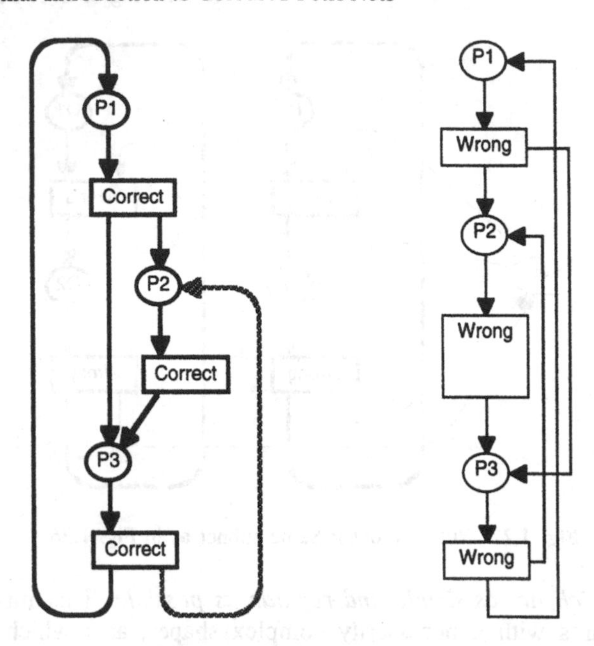

Fig. 1.29. Rule 6: Avoid parallel arcs next to each other

7. *Draw dashed lines correctly.* The rule is illustrated by means of arcs, but it applies to all kinds of lines, including the outline of nodes. It is sometimes a good idea to use dashed lines, in order to distinguish certain lines from other lines. However, it is a very common error to use dashes that are too short, have too much space in between, or by-pass the corners. This makes it unnecessarily difficult to follow the lines.

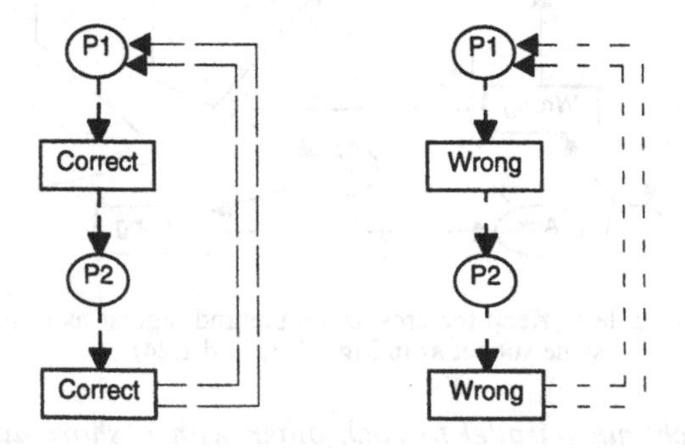

Fig. 1.30. Rule 7: Draw dashed lines correctly

8. *Avoid shapes which can be confused with each other.* Some modellers use rounded boxes for places (instead of ellipses). This makes it possible to have more text inside the places, but it makes it too difficult to distinguish between places and transitions. There is simply too little difference between the two shapes. Moreover, we may sometimes have a set of crossing arcs forming a

shape which may be confused with a transition. This is illustrated by the area marked "NoTrans", in the rightmost subnet of Fig. 1.31.

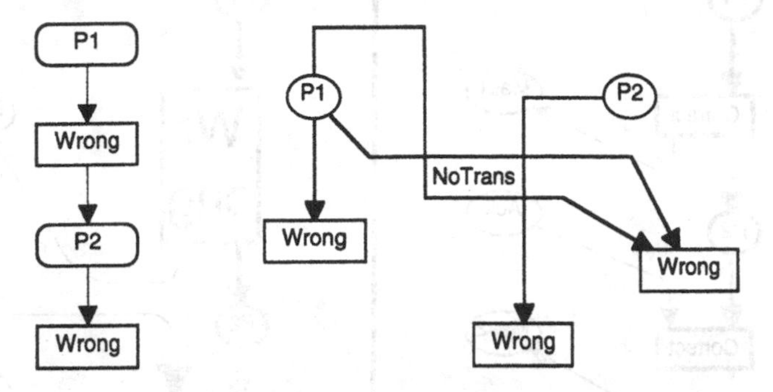

Fig. 1.31. Rule 8: Avoid shapes which can be confused with each other

9. Position the net inscriptions as close to the corresponding places, transitions and arcs as possible. This rule is in Fig. 1.32 illustrated by means of arc expressions, but it applies to all kinds of net inscriptions. When the arcs have different colours, it is a good idea to give each arc expression the same colour as the corresponding arc.

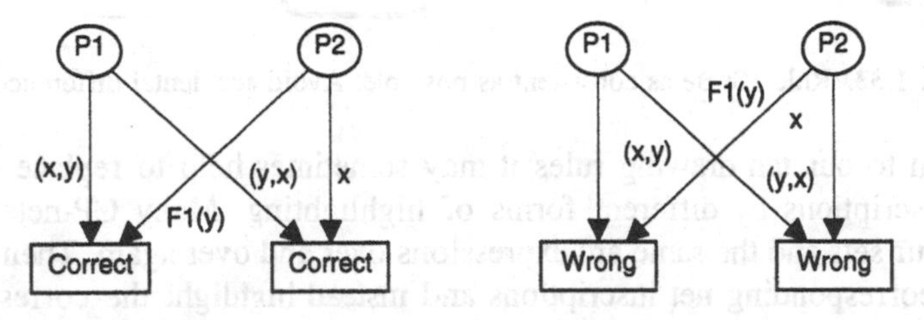

Fig. 1.32. Rule 9: Position the net inscriptions very close to the net elements

10. Try to be as consistent as possible. If we want to be able to show different relationships by highlighting the corresponding net elements, it is important that this attempt is not drowned by a number of accidental differences. Each of the two subnets in Fig. 1.33 highlights a cyclic subnet by using a different line thickness, and by positioning the net elements on a figure with a nice geometrical shape. In the leftmost subnet this works fine, but in the rightmost it is blurred by all the many accidental differences in size, arc shape, line shading, etc. In principle a modeller should always be able to tell why he has drawn two graphical objects in a different way. Otherwise he should draw them to be as similar as possible.

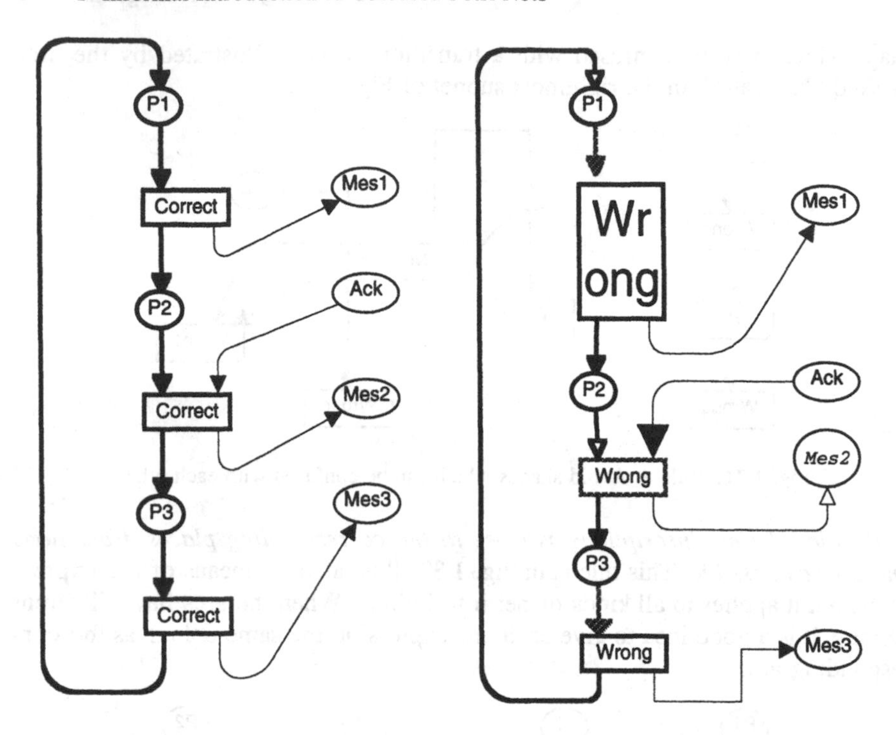

Fig. 1.33. Rule 10: Be as consistent as possible. Avoid accidental differences

In addition to our ten drawing rules it may sometimes help to replace some of the net inscriptions by different forms of highlighting. Many CP-nets use the same colour sets and the same arc expressions over and over again. Then we may omit the corresponding net inscriptions and instead highlight the corresponding places in different ways – as illustrated by Figs. 1.34 and 1.35. Notice the legend in Fig. 1.35 saying how the different kinds of highlighting should be interpreted. Using a computer tool we would not omit the net inscriptions, but we would make them invisible. The legend would then consist of auxiliary objects, and it would not influence the behaviour – which would be determined from the net inscriptions (visible or not).

Making a readable CP-net is a bit of an art, and it often takes a considerable amount of patience and experiment before a good solution is found. Experience means a lot for the speed at which a readable version of a given CP-net can be constructed. However, nowadays the task is much easier – even for the novice – due to the availability of graphical Petri net editors. It is now easy and fast to try several different layouts, without having to redraw the entire net or use large amounts of correction tape/fluid. All the figures in this book were drawn using the CPN editor described in Chap. 6. All the nets in Sect. 1.6 were drawn in approximately 2 hours.

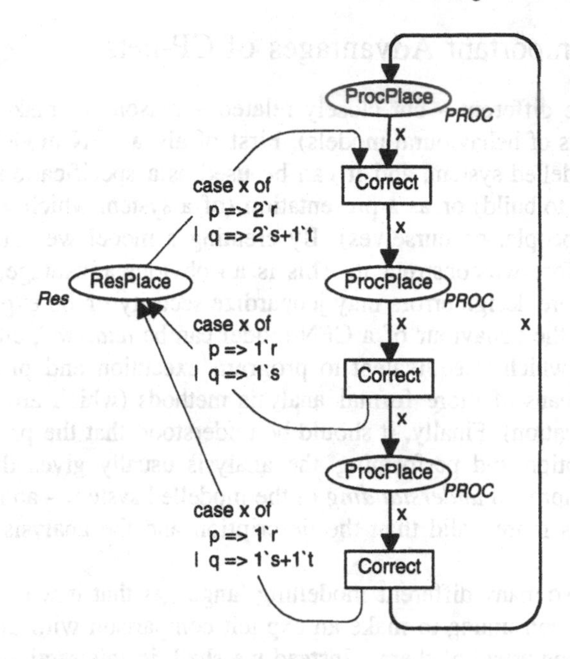

Fig. 1.34. CP-net with many identical colour set inscriptions and arc expressions

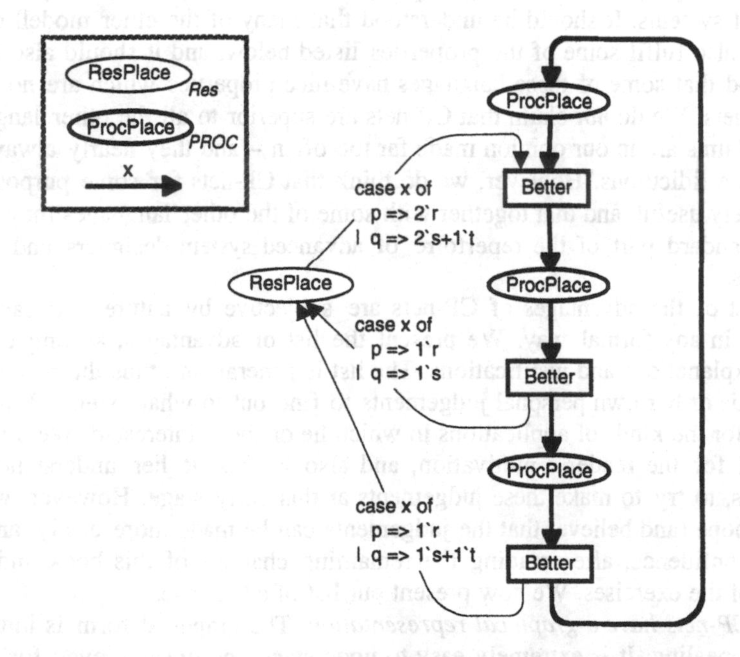

Fig. 1.35. Same CP-net as in Fig. 1.34, but now we have omitted some of the net inscriptions and instead highlighted the corresponding net elements in different ways

1.7 Some Important Advantages of CP-nets

There are three different – but closely related – reasons to make CPN models (and other kinds of behavioural models). First of all, a CPN model is a *description* of the modelled system, and it can be used as a specification (of a system which we want to build) or as a presentation (of a system which we want to explain to other people, or ourselves). By creating a model we can investigate a new system before we construct it. This is an obvious advantage, in particular for systems where design errors may jeopardize security or be expensive to correct. Secondly, the behaviour of a CPN model can be *analysed*, either by means of simulation (which is equivalent to program execution and program debugging) or by means of more formal analysis methods (which are equivalent to program verification). Finally, it should be understood that the process of creating the description and performing the analysis usually gives the modeller a dramatically improved *understanding* of the modelled system – and it is often the case that this is more valid than the description and the analysis results themselves.

There exist so many different modelling languages that it would be very difficult and time consuming to make an explicit comparison with all of them (or even the most important of them). Instead we shall, in this section, make an implicit comparison, by listing twelve of those properties which make CP-nets a valuable language for the design, specification and analysis of many different types of systems. It should be understood that many of the other modelling languages also fulfil some of the properties listed below, and it should also be understood that some of these languages have nice properties which are not found in CP-nets. We do *not* claim that CP-nets are superior to all the other languages. Such claims are in our opinion made far too often – and they nearly always turn out to be ridiculous. However, we do think that CP-nets for some purposes are extremely useful, and that together with some of the other languages they should be a standard part of the repertoire of advanced system designers and system analysts.

Most of the advantages of CP-nets are subjective by nature and cannot be proved in any formal way. We present the list of advantages, accompanied by brief explanations and justifications. The list is general, and thus the reader must make his or her own personal judgements to find out to what extent CP-nets are useful for the kinds of applications in which he or she is interested. We think it is fruitful for the reader's motivation, and also for his or her understanding of CP-nets, to try to make these judgements at this early stage. However, we certainly hope (and believe) that the judgements can be made more easily, and with more confidence, after reading the remaining chapters of this book and doing some of the exercises. We now present our list of advantages.

1. CP-nets have a graphical representation. The graphical form is intuitively very appealing. It is extremely easy to understand and grasp – even for people who are not very familiar with the details of CP-nets. This is due to the fact that CPN diagrams resemble many of the informal drawings which designers and engineers make while they construct and analyse a system. Just think about how often you have illustrated an algorithm or a communication protocol by drawing a directed graph, where the nodes represent states and actions, while the arcs describe how to go from one state to another, by executing some of the actions. The notions of states, actions and flow are basic to many kinds of system and these concepts are – in a very vivid and straightforward way – represented by the places, transitions and arcs of CP-nets.

2. *CP-nets have a well-defined semantics which unambiguously defines the behaviour of each CP-net.* It is the presence of the semantics which makes it possible to implement simulators for CP-nets, and it is also the semantics which forms the foundation for the formal analysis methods described in Chap. 5 and in Vol. 2.

3. *CP-nets are very general and can be used to describe a large variety of different systems.* The applications of CP-nets range from informal systems (such as the description of work processes) to formal systems (such as communication protocols). They also range from software systems (such as distributed algorithms) to hardware systems (such as VLSI chips). Finally, they range from systems with a lot of concurrent processes (such as flexible manufacturing) to systems with no concurrency (such as sequential algorithms).

4. *CP-nets have very few, but powerful, primitives.* The definition of CP-nets is rather short and it builds upon standard concepts which many system modellers already know from mathematics and programming languages. This means that it is relatively easy to learn to use CP-nets. However, the small number of primitives also means that it is much easier to develop strong analysis methods.

5. *CP-nets have an explicit description of both states and actions.* This is in contrast to most system description languages which describe either the states or the actions – but not both. Using CP-nets, the reader may easily change the point of focus during the work. At some instances of time it may be convenient to concentrate on the states (and almost forget about the actions) while at other instances it may be more convenient to concentrate on the actions (and almost forget about the states).

6. *CP-nets have a semantics which builds upon true concurrency, instead of interleaving.* This means that the notions of conflict, concurrency and causal dependency can be defined in a very natural and straightforward way (as we have seen in Sect. 1.3). In an interleaving semantics it is impossible to have two actions in the same step, and thus concurrency only means that the actions can occur after each other, in any order. In our opinion, a true-concurrency semantics is easier to work with – because it is closer to the way human beings usually think about concurrent actions.

7. *CP-nets offer hierarchical descriptions.* This means that we can construct a large CP-net by relating smaller CP-nets to each other, in a well-defined way. The hierarchy constructs of CP-nets play a role similar to that of subroutines, procedures and modules of programming languages, and it is the existence of hierarchical CP-nets which makes it possible to model very large systems in a manageable and modular way. Hierarchical CP-nets will be introduced in Chap. 3.

8. CP-nets integrate the description of control and synchronization with the description of data manipulation. This means that on a single sheet of paper it can be seen what the environment, enabling conditions and effects of an action are. Many other graphical description languages work with graphs which only describe the environment of an action – while the detailed behaviour is specified separately (often by means of unstructured prose).

9. CP-nets are stable towards minor changes of the modelled system. This is proved by many practical experiences and it means that small modifications of the modelled system do not completely change the structure of the CP-net. In particular, it should be observed that this is also true when a number of subnets describing different sequential processes are combined into a larger CP-net. In many other description languages, e.g., finite automata, such a combination often yields a description which it is difficult to relate to the original sub-descriptions.

10. CP-nets offer interactive simulations where the results are presented directly on the CPN diagram. The simulation makes it possible to debug a large model while it is being constructed – analogously to a good programmer debugging the individual parts of a program as he finishes them. The colours of the moving tokens can be inspected.

11. CP-nets have a large number of formal analysis methods by which properties of CP-nets can be proved. There are four basic classes of formal analysis methods: construction of occurrence graphs (representing all reachable markings), calculation and interpretation of system invariants (called place and transition invariants), reductions (which shrink the net without changing a certain selected set of properties) and checking of structural properties (which guarantee certain behavioural properties).

12. CP-nets have computer tools supporting their drawing, simulation and formal analysis. This makes it possible to handle even large nets without drowning in details and without making trivial calculation errors. The existence of such computer tools is extremely important for the practical use of CP-nets.

In this section we have listed a number of advantages of CP-nets. Many of these are also valid for other kinds of high-level nets, PT-nets, and other kinds of modelling languages. Once more, we want to stress that we do not view CP-nets as "the superior" system description language. In contrast, we consider the world of computer science to be far too complicated and versatile to be handled by a single language. Thus we think CP-nets must be used together with many other kinds of modelling languages. Often it is valuable to use different languages to describe different aspects of the system, and then the resulting set of descriptions should be considered as complementary, not as alternatives.

Bibliographical Remarks

The foundation of Petri nets was presented by Carl Adam Petri in his doctoral thesis [85]. The first nets were called Condition/Event Nets (CE-nets). This net model allows each place to contain at most one token – because the place is considered to represent a boolean condition, which can be either true or false. In the following years a large number of people contributed to the development of new net models, basic concepts, and analysis methods. One of the most notable results was the development of Place/Transition Nets (PT-nets). This net model allows a place to contain several tokens. The first coherent presentation of the theory and application of Petri nets was given in the course material developed for the *First Advanced Course on Petri Nets* in 1979 [14] and this was later supplemented by the course material for the *Second Advanced Course on Petri Nets* in 1986 [15] and [16]. In addition there exist a number of text books on PT-nets, e.g., [8], [84], [92], [96] (in English), [6], [93], [106] (in German), [11] (in French), [12] (in Italian) and [13], [103] (in Spanish). A short introduction to Petri nets is given in [76]. A full Petri net bibliography can be found in [88], which contains more than 4000 entries. Updated versions of the bibliography are published at regular intervals in *Advances in Petri Nets.*

For theoretical considerations CE-nets turned out to be more tractable than PT-nets, and much of the theoretical work concerning the definition of basic concepts and analysis methods has been performed on CE-nets. Later, a new net model called Elementary Nets (EN-nets) was proposed in [97] and [108]. The basic ideas of this net model are very close to those of CE-nets – but EN-nets avoid some of the technical problems which turned out to be present in the original definition of CE-nets.

For practical applications, PT-nets were used. However, it often turned out that this net model was too low-level to cope with the real-world applications in a manageable way, and different researchers started to develop their own extensions of PT-nets – adding concepts such as: priority between transitions, time delays, global variables to be tested and updated by transitions, zero testing of places, etc. In this way a large number of different net models were defined. However, most of these net models were designed with a single, and often very narrow, application area in mind. This created a serious problem. Although some of the net models could be used to give adequate descriptions of certain systems, most of the net models possessed almost no analytic power. The main reason for this was the large variety of different net models. It often turned out to be a difficult task to translate an analysis method developed for one net model to another – and in this way the efforts to develop suitable analysis methods were widely scattered.

The breakthrough with respect to this problem came when Predicate/Transition Nets (PrT-nets) were presented in [39]. PrT-nets were the first kind of high-level nets which were constructed without any particular application area in mind. PrT-nets form a nice generalization of PT-nets and CE-nets (exploiting the same kind of reasoning that leads from propositional logic to predicate logic). PrT-nets can be related to PT-nets and CE-nets in a formal way – and this makes it possible to generalize most of the basic concepts and analysis methods that have been developed for these net models – so that they also become applicable to PrT-nets. Later, an improved definition of PrT-nets was presented in [41]. This definition draws heavily on sigma algebras (as known from the theory of abstract data types).

However, it soon became apparent that PrT-nets present some technical problems when the analysis methods of place invariants and transition invariants are generalized. It is possible to calculate invariants for PrT-nets, but the interpretation of the invariants is difficult and must be done with great care to avoid erroneous results. The problem arises because of the variables which appear in the arc expressions of PrT-nets. These variables also appear in the invariants, and to interpret the invariants it is necessary to bind the variables, via a complex set of substitution rules. To overcome this problem the first version of Coloured Petri Nets (CP^{81}-nets) was defined in [51]. The main ideas of this net model are directly inspired by PrT-nets, but the relation between a binding element and the token colours involved in the occurrence is now defined by functions and not by expressions as in PrT-nets. This removes the variables, and invariants can be interpreted without problems.

However, it often turns out that the functions attached to arcs in CP^{81}-nets are more difficult to read and understand than the expressions attached to arcs in PrT-nets. Moreover, as indicated above, there is a strong relation between PrT-nets and CP^{81}-nets – and from the very beginning it was clear that most descriptions in one of the net models could be informally translated to the other net model, and vice versa. This lead to the idea of an improved net model – combining the qualities of PrT-nets and CP^{81}-nets. This net model was defined in [52] where the nets were called High-level Petri Nets (HL-nets). Unfortunately, this name has given rise to a lot of confusion since the term "high-level nets" at that time started to become used as a generic name for PrT-nets, CP^{81}-nets, HL-nets, and several other kinds of net models. To avoid this confusion it was necessary to change the name from HL-nets to Coloured Petri Nets (CP-nets).

CP-nets have two different representations. The *expression representation* uses arc expressions and guards, while the *function representation* uses linear functions between multi-sets. Moreover, there are formal translations between the two representations (in both directions). The expression representation is nearly identical to PrT-nets (as presented in [39]), while the function representation is nearly identical to CP^{81}-nets. The first coherent presentations of CP-nets and their analysis methods were given in [54] and [58]. In [52] and [54] we used the expression representation to describe systems, while we used the function representation for all the different kinds of analysis. However, it has turned out that it only is necessary to turn to linear functions when we deal with invariants analysis, and this means that in [58] and this book we use the expression representation for all purposes – except for the calculation of invariants. This change is important for the practical use of CP-nets, because it means that the function representation and the translations (which are mathematically rather complicated) are no longer part of the basic definition of CP-nets. Instead they are part of the invariant method (which anyway demands considerable mathematical skills).

Today most of the practical applications of Petri nets (reported in the literature) use either PrT-nets or CP-nets – although several other kinds of high-level nets have been proposed. There is very little difference between PrT-nets and CP-nets (and many modellers do not make a clear distinction between the two kinds of net models). The main differences between the two net models are today hidden inside the methods to calculate and interpret place and transition invariants (and this is of course not surprising when you think about the original motivation behind the development of CP[81]-nets). Instead of viewing PrT-nets and CP-nets as two different modelling languages it is, in our opinion, much more adequate to view them as two slightly different dialects of the same language.

As mentioned above, several other classes of high-level nets have been defined, e.g., algebraic nets [30], CP-nets with algebraic specifications [116], many-sorted high-level nets [9], numerical Petri nets [10], [107], OBJSA nets [5], PrE-nets with algebraic specifications [65], Petri nets with structured tokens [95] and relation nets [94]. All these net classes are quite similar to CP-nets, but use different inscription languages (e.g., building on algebraic specifications or object oriented languages). Some of the most important papers on high-level nets, their analysis methods and applications have been reprinted in [59].

The functional programming language Standard ML has been developed at Edinburgh University. It is used for the inscriptions of CP-nets, and it is also one of the programming languages used in the implementation of the CPN tools described in Chap. 6. Standard ML has been described in a number of papers and text books – among them [45], [73], [74], [82], [91], [105], [109] and [119].

The drawing rules presented in Sect. 1.6 are – to a very large extent – based on the work of Horst Oberquelle, [80].

Exercises

Exercise 1.1.

For scheduling purposes it is usual to describe a work process as a set of partially related tasks. As a very simple (and naive) example, the construction of a house may be described in the following way, where an arc from one task to another indicates that the first task must be fully completed before the second can be started.

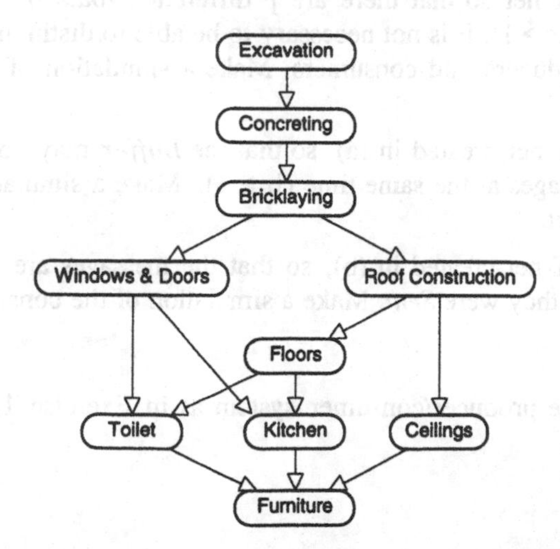

(a) Modify the diagram above so that it becomes a PT-net, where each task is represented by a transition (while places represent the states which exist between the ending of some tasks and the beginning of the succeeding ones). Make a simulation of the constructed PT-net.

(b) Modify the diagram above so that it becomes a PT-net, where each task is represented by a place (while transitions represent the ending of some tasks and the beginning of the following ones). Make a simulation of the constructed PT-net.

(c) Discuss the two PT-nets constructed above. Which one is the more suitable – in the sense that you do not have to introduce places or transitions without a straightforward interpretation in the original system?

Exercise 1.2.
The PT-net below describes a *Producer* who *Produces* and *Sends* some kind of messages via a *Buffer* to a *Consumer* who *Receives* and *Consumes* them.

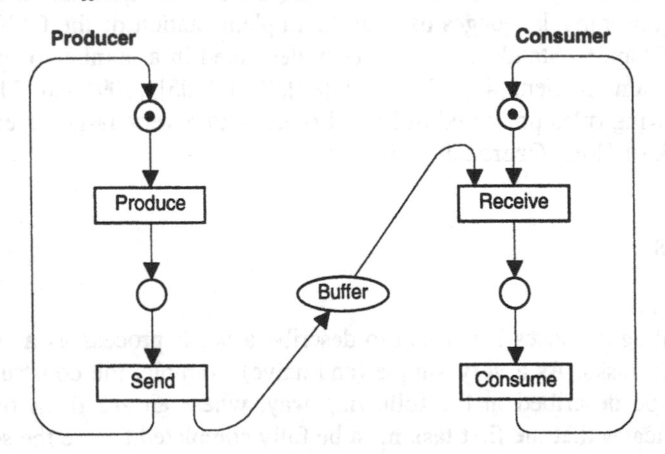

(a) Modify the PT-net so that there are p different *Producers* and c different *Consumers* ($p,c \geq 1$). It is not necessary to be able to distinguish between the individual producers and consumers. Make a simulation of the constructed PT-net.

(b) Modify the PT-net created in (a), so that the *Buffer* may contain at most b different messages at the same time ($b \geq 1$). Make a simulation of the constructed PT-net.

(c) Modify the PT-net created in (b), so that the messages are *Received* in the same order as they were *Sent*. Make a simulation of the constructed PT-net.

Exercise 1.3.
Consider the same producer/consumer system as in Exercise 1.2. Assume that there is:

- A set of *Producers:* PROD = {$P_1, P_2, \ldots, P_{np}$} with np ≥ 1.
- A set of *Consumers:* CONS = {$C_1, C_2, \ldots, C_{nc}$} with nc ≥ 1.
- A set of *Data:* DATA = {$D_1, D_2, \ldots, D_{nd}$} with nd ≥ 1.

The system may be described by the CP-net below (in which we have 3 producers, 2 consumers and 4 different kinds of data). Notice that the variable, d (on the output arc of *Send),* does not appear in the guard or input arc expression of *Send*. This means that the binding of d does not influence the enabling of *Send*, and thus each producer can send all kinds of messages (from DATA).

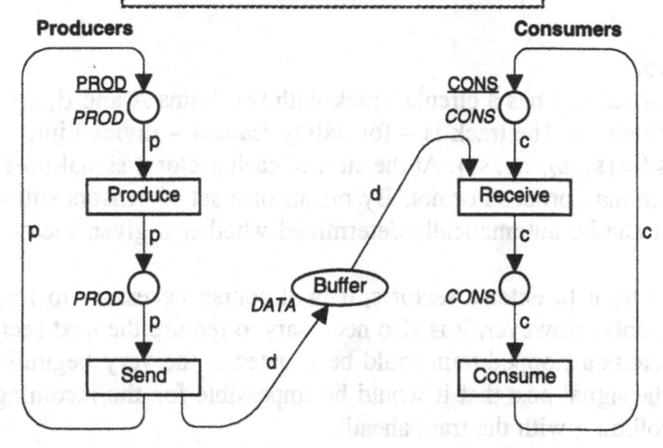

(a) Modify the CP-net so that each message is *Sent* to a single specified consumer (who is the only one who may *Receive* it). This can be done by changing the colour set for the *Buffer* from DATA to CONS×DATA (where CONS×DATA denotes the cartesian product of CONS and DATA, i.e., all pairs (c,d) such that c∈CONS and d∈DATA). Make a simulation of the constructed CP-net.

(b) Modify the CP-net created in (a), so that the messages are *Received* in the same order as they were *Sent*. This can be done by changing the colour set for the *Buffer* to (CONS×DATA)* containing all finite lists with elements from CONS×DATA. Make a simulation of the constructed CP-net.

Exercise 1.4.

A semaphore is a device which allows two or more processes to synchronize their actions, e.g., to access shared memory in a safe way. A semaphore has two operations, V and P (also called signal and wait). Each of these operations updates the semaphore value, which is of type integer. The V operation increases the semaphore by one, while the P operation decreases it by one. However, it is guaranteed that the semaphore value never becomes negative. This implies that it may be necessary to postpone the execution of a P operation until more V operations have occurred.

(a) Construct a PT-net modelling a semaphore system. The net should have two transitions (modelling the V and P operations) and a single place (modelling the semaphore value). Make a simulation of the constructed PT-net. Can the V and P transitions be concurrent to each other? Can each of them be concurrent to itself?

(b) Construct a CP-net modelling a semaphore system. Again, the net should have two transitions (modelling the V and P operations) and a single place (modelling the semaphore value). The place should have a colour set which is of type integer. Make a simulation of the constructed CP-net. Can the V and P transitions be concurrent to each other? Can each of them be concurrent to itself?

Exercise 1.5.
A small model railway has a circular track with two trains A and B, which move in the same direction. The track is – for safety reasons – divided into seven different sectors $S=\{s_1, s_2, \ldots, s_7\}$. At the start of each sector a signal post indicates whether a train may proceed or not. By means of a set of sensors situated at the signal posts it can be automatically determined whether a given sector is empty or not.

To allow a train to enter a sector s_i it is of course necessary to require that sector to be empty. However, it is also necessary to require the next sector s_{i+1} to be empty (because a stopped train could be situated at the very beginning of s_{i+1} – so near to the signal post that it would be impossible for the incoming train to stop before colliding with the train ahead).

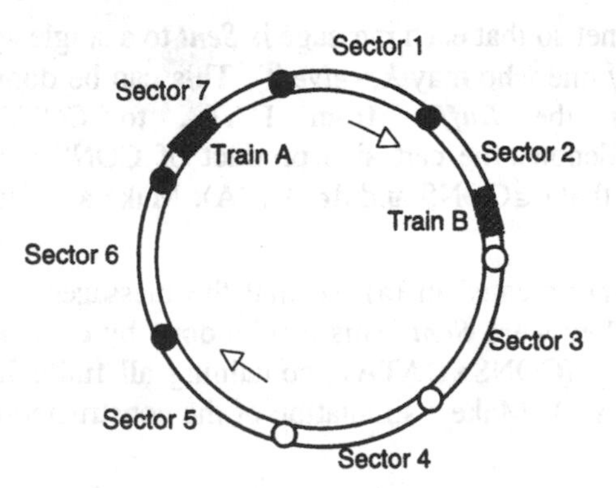

(a) Describe the train system by a PT-net. Each sector s_i may be represented by three places O_{ia}, O_{ib}, and E_i (where O_{ix} is shorthand for "sector s_i occupied by train x" and E_i is shorthand for "sector s_i empty"). Make a simulation of the constructed PT-net.

(b) Describe the same system by a CP-net where each sector is described by two places O_i and E_i. Then O_i has the set $\{a,b\}$ as possible token colours, while E_i

has E={e} as token colours. Make a simulation of the constructed CP-net. Compare the CP-net with the PT-net from (a).

(c) Describe the same system by a CP-net which only has two places O and E, and a single transition *Move to Next Sector*. Make a simulation of the constructed CP-net.

Exercise 1.6.
Imagine a system where five Chinese philosophers are situated around a circular table. In the middle of the table there is a delicious dish of rice, and between each pair of philosophers there is a single chopstick. Each philosopher alternates between thinking and eating. To eat, the philosopher needs two chopsticks, and he is only allowed to use the two which are situated next to him (on his left and right side). It is obvious that this restriction (lack of resources) prevents two neighbours from eating at the same time.

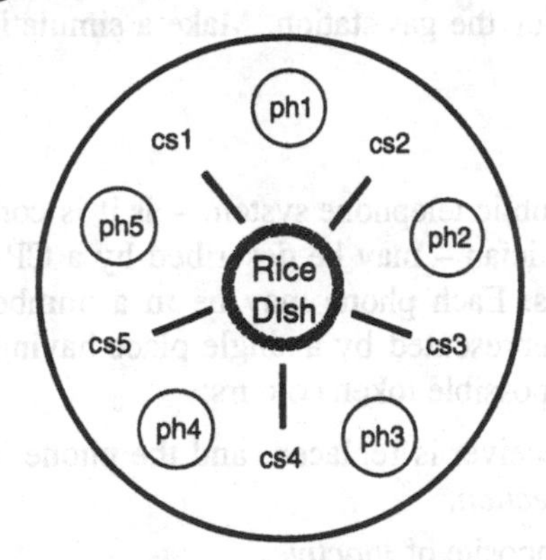

(a) Describe the philosopher system by a PT-net. It is assumed that each philosopher simultaneously (and indivisibly) picks up his pair of chopsticks. Analogously, he puts them down in a single indivisible action. Each philosopher may be represented by two places (*Think* and *Eat*) and two transitions (*Take Chopsticks* and *Put Down Chopsticks*). Each chopstick may be represented by a single place (which has a token when the chopstick is unused). Make a simulation of the constructed PT-net.

(b) Describe the same system by a CP-net which contains the two colour sets PH = $\{ph_1, ph_2, \ldots, ph_5\}$ and CS = $\{cs_1, cs_2, \ldots, cs_5\}$, representing the philosophers and the chopsticks, respectively. It is only necessary to use three places (*Think*, *Eat* and *Unused Chopsticks*) and two transitions (*Take Chopsticks* and *Put Down Chopsticks*). Make a simulation of the constructed CP-net.

(c) Modify the CP-net created in (b), so that each philosopher first takes his right chopstick and next the left one. Analogously, he first puts down his left chopstick and next the right one. Make a simulation of the constructed CP-net. Does this modification change the overall behaviour of the system (e.g., with respect to deadlocks and fairness between the philosophers)?

(d) Modify the CP-net created in (c), so that each philosopher takes the two chopsticks one at a time (but in an arbitrary order, which may change from one eating time to the next). Analogously, he puts down the two chopsticks one at a time (in an arbitrary order). Make a simulation of the constructed CP-net. Does this modification change the overall behaviour of the system (e.g., with respect to deadlocks and fairness between the philosophers)?

Exercise 1.7.
Consider a gas station with a number of customers, three different pumps and one operator. When a customer arrives at the gas station he pays the operator. Then the operator activates the relevant pump. The customer pumps the desired quantity of gasoline (the maximum quantity is of course determined by the pre-paid amount of money). Finally, the customer receives his change from the operator (if any).

(a) Construct a CP-net for the gas station. Make a simulation of the constructed CP-net.

Exercise 1.8.
The functioning of the public telephone system – as it is conceived by a user and not by a telephone technician – may be described by a CP-net with less than 15 places and 15 transitions. Each phone may be in a number of different states, each of which may be represented by a single place having the set of all phone numbers U as the set of possible token colours:

Inactive — the receiver is replaced, and the phone is not involved in any *Connection*,

Engaged — the opposite of *Inactive*,

Continuous — the receiver has been lifted and a continuous tone indicates that a number may now be dialled,

No Tone — for a short period, while it is investigated whether the dialled phone is *Inactive* or *Engaged*,

Short — tone with short intervals, indicating that the dialled phone is *Engaged*,

Long — tone with long intervals, indicating that the dialled phone is *Inactive*,

Ringing — the phone is ringing,

Connected — the phone is connected to another phone, and a conversation may take place.

In addition it is necessary to have three places representing the state of the telephone exchange. Each of these places has U×U as possible token colours. This means that each token is a pair (x,y), where x∈U is the dialling phone and y∈U is the dialled phone:

Request — from phone x the number y has been dialled (but nothing else has happened yet),

Call	the phone y turned out to be *Inactive* – now x has a tone with *Long* intervals, while y is *Ringing*,
Connection	the two phones x and y are *Connected*.

The part of the CP-net which describes the dialling of a number may look as shown below.

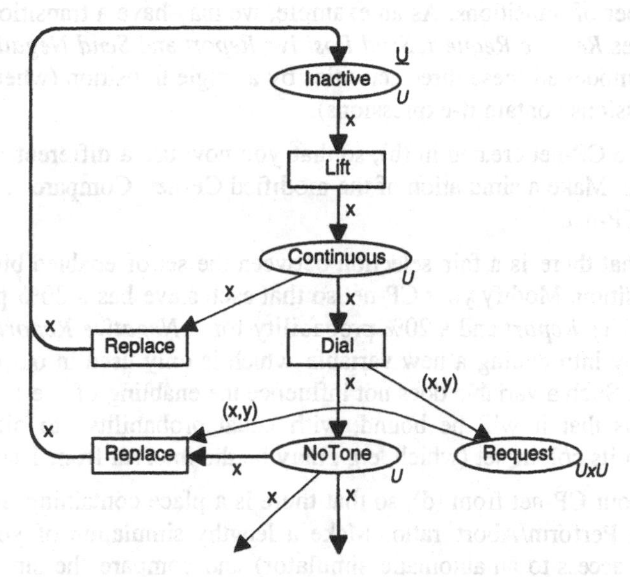

(a) Extend the CP-net so that it becomes a complete description of the public telephone service. Ignore time-outs, local switchboards, conference calls, etc.

(b) Make a simulation of the constructed CP-net. What happens when a phone dials its own number? Can the dialled phone break the connection, or is it only the dialling phone? Is this a correct model of the telephone system in your country? If not, how should it be modified?

(c) Modify the CP-net so that there are unused phone numbers. An unused number can be activated at any time, while a used number only can be deactivated when the corresponding phone is *Inactive*.

(d) Modify the CP-net so that it also describes time-outs.

Exercise 1.9.

Consider a small cyclic system in which a single *Master* process communicates with a number of identical *Slave* processes. The master *Sends a Request* to all the slaves. Each slave *Receives* the Request, *Checks* whether the request can be fulfilled and *Reports* this to the master – but the slave does *not* perform the request yet. The master *Receives the Reports* and iff all reports are positive the master *Informs* the slaves that they are allowed to *Perform* the task. Otherwise they are told to *Abort* the Task.

(a) Construct a CP-net for the Master/Slave system. Make a simulation of the constructed CP-net.

(b) Is the master able to make a new request before all the Slaves have returned to their initial state? If not, modify your CP-net to allow this, without mixing the handling of two different requests. Make a simulation of the modified CP-net.

For the Master/Slave system, and many other systems, it is possible to use a varying number of transitions. As an example, we may have a transition for each of the activities *Receive Request*, *Send Positive Report* and *Send Negative Report* – or we may model all these three activities by a single transition (where some of the arc expressions contain if-expressions).

(c) Modify the CP-net created in (b), so that you now use a different number of transitions. Make a simulation of the modified CP-net. Compare it with your original CP-net.

(d) Assume that there is a fair selection between the set of enabled bindings for each transition. Modify your CP-net so that each slave has a 80% probability for a *Positive Report* and a 20% probability for a *Negative Report*. This can be done by introducing a new variable which is only used in output arc expressions. Such a variable does not influence the enabling of the transitions – this means that it will be bound, with equal probability, to all different colours in its colour set (which, e.g., may be the interval from 1 and 10).

(e) Modify your CP-net from (d), so that there is a place containing information about the Perform/Abort ratio. Make a lengthy simulation of your net (if you have access to an automatic simulator) and compare the simulation results with the theoretical Perform/Abort ratio (which it is easy to determine).

Exercise 1.10.

The CP-net below violates many of the drawing rules from Sect. 1.5.

(a) Modify the CP-net so that the drawing rules, as far as possible, are followed. Try to make different alternative layouts and different highlightings, etc.

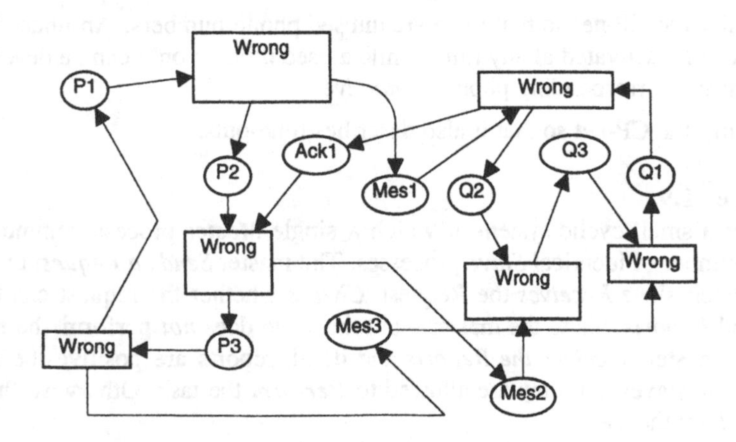

Exercise 1.11.
Take two or three of your old CP-nets – or some nets of a friend.

(a) Check whether the CP-nets fulfil the drawing rules and naming/comment guidelines from Sect. 1.6. This should be done by checking each individual rule/guideline against your CP-nets.

(b) Modify the CP-nets so that the drawing rules and naming/comment guidelines are followed – as far as possible. Try to make different alternative layouts and different highlightings, etc.

Chapter 2

Formal Definition of Coloured Petri Nets

This chapter contains the formal definition of non-hierarchical CP-nets and their behaviour. A non-hierarchical CP-net is defined as a many-tuple. However, it should be understood that the only purpose of this is to give a mathematically sound and unambiguous definition of CP-nets and their semantics. Any concrete net, created by a modeller, will always be specified in terms of a CPN diagram (i.e., a diagram similar to Figs. 1.7 and 1.13) – or by a hierarchical CPN diagram (to be introduced in Chap. 3). It is in principle (but not in practice) easy to translate a CPN diagram into a CP-net tuple, and vice versa. The tuple form is adequate when we want to formulate general definitions and prove theorems which apply to all (or a large class) of CP-nets. The graph form is adequate when we want to construct a particular CP-net modelling a specific system.

Also we investigate the relationship between non-hierarchical CP-nets and Place/Transition Nets (PT-nets), and it turns out that each CP-net can be translated into a behaviourally equivalent PT-net, and vice versa. The translation from CP-nets to PT-nets is unique, while the translation in the other direction can be done in many different ways. The existence of an equivalent PT-net is extremely useful, because it tells us how to generalize the basic concepts and the analysis methods of PT-nets to non-hierarchical CP-nets. We simply define these concepts in such a way that a CP-net has a given property iff the equivalent PT-net has the corresponding property.

Section 2.1 contains the formal definition of multi-sets and their operations. Section 2.2 defines the structure of non-hierarchical CP-nets and it introduces some notation to discuss the relationship between neighbouring elements of the net structure. Section 2.3 defines the behaviour of non-hierarchical CP-nets, i.e., markings, steps, enabling and occurrence sequences. Finally, Sect. 2.4 investigates the relationship between non-hierarchical CP-nets and PT-nets; the section is theoretical and can be skipped by readers who are interested primarily in the practical application of CP-nets.

2.1 Multi-sets

Intuitively, a multi-set is the same as a set – except that there may be **multiple appearances** of the same element. If we have a set {b,c,f} and add the element c we still have the set {b,c,f}. However, if we have the multi-set {b,c,f} and add the element c we get the multi-set {b,c,c,f}, which now has two appearances of the element c. A multi-set is also sometimes called a *bag*, because it may contain multiple appearances – in a way similar to which a bag may contain identical items. As shown in Fig. 2.1, it is possible to define different operations on multi-sets. All these operations are generalizations of the corresponding operations defined for sets.

A multi-set is always defined **over a set S**, and this means that the elements of the multi-set are taken from S. To determine the multi-set it is sufficient to know, for each element in S, how many times that element appears in the multi-set. This can be described by a formal sum where each element of S has a coefficient saying how many times it appears, or by a function mapping the elements of S into the non-negative integers. As an example, we consider the two multi-sets m_1 and m_2 from Fig. 2.1. They are multi-sets over the set S = {a,b,c,d,e} and they are represented by the following formal sums – where by convention we omit elements with zero coefficients:

$$m_1 = 1`a+2`c+1`e$$
$$m_2 = 1`a+2`b+3`c+1`e.$$

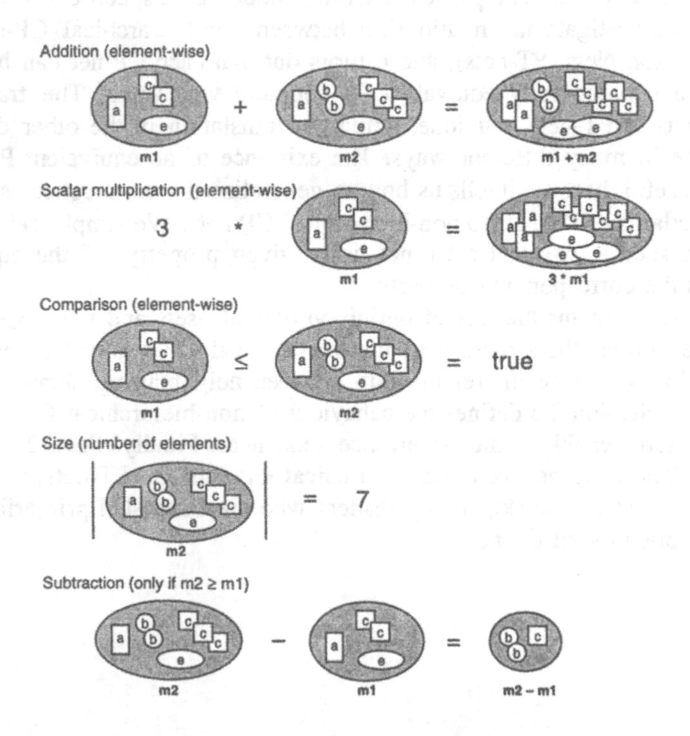

Fig. 2.1. Some operations on multi-sets

Notice that (in contrast to standard mathematics) we use an explicit operator between the coefficients and the elements and include coefficients which are equal to one. These conventions make it easier to deal with multi-sets over integers, e.g., 3`1+2`35+1`59 (and they also make it easier to perform automatic type checking of multi-sets in CPN ML). As mentioned above, it is also possible to represent the two multi-sets m_1 and m_2 as functions:

$$m_1(s) = \begin{cases} 1 & \text{if } s \in \{a,e\} \\ 2 & \text{if } s=c \\ 0 & \text{otherwise,} \end{cases}$$

$$m_2(s) = \begin{cases} 1 & \text{if } s \in \{a,e\} \\ 2 & \text{if } s=b \\ 3 & \text{if } s=c \\ 0 & \text{otherwise.} \end{cases}$$

Now we can give the formal definition of multi-sets. We use $\mathbb{N}$ to denote the set of all non-negative integers and we use $[A \to B]$ to denote the set of all functions from A to B.

Definition 2.1: A **multi-set** m, over a non-empty set S, is a function $m \in [S \to \mathbb{N}]$. The non-negative integer $m(s) \in \mathbb{N}$ is the **number of appearances** of the element s in the multi-set m.

We usually represent the multi-set m by a formal sum:

$$\sum_{s \in S} m(s)`s.$$

By S_{MS} we denote the set of all multi-sets over S. The non-negative integers $\{m(s) \mid s \in S\}$ are called the **coefficients** of the multi-set m, and m(s) is called the **coefficient** of s. An element $s \in S$ is said to **belong** to the multi-set m iff $m(s) \neq 0$ and we then write $s \in m$.

We shall use Ø to denote the empty multi-set, i.e., the multi-set where each element has zero as its number of appearances. In CPN ML the empty multi-set is denoted by "empty" (since "Ø" is not an ASCII symbol). To be precise there is an empty multi-set for each element set S. However, we shall ignore this and allow ourselves to speak about *the* empty multi-set – in a similar way that we speak about *the* empty set and *the* empty list.

The set of all subsets of S will be denoted S_S. Obviously, there is a one-to-one correspondence between S_S and those multi-sets $m \in S_{MS}$ where $0 \leq m(s) \leq 1$ for all $s \in S$. Thus we shall, without any further comments, allow ourselves to consider each set as a multi-set (and each multi-set with $0 \leq m(s) \leq 1$ as a set). This means that for each set $A \subseteq S$ we use A to denote the multi-set which corresponds to the set A, i.e., the multi-set that has exactly one appearance of each element in A (and no other elements). Analogously, for each element $s \in S$ we use "s" to denote the multi-set that corresponds to $\{s\}$, i.e., the multi-set 1`s.

The operations which we illustrated in Fig. 2.1 can now be defined as shown in Def. 2.2. It should be recalled that multi-sets are functions into the non-negative integers, and it can then be seen that the operations (with the exception of $| \, |$) are nothing other than the standard operations on functions (which have a range where +, –, *, and the comparison operators =, ≠, ≤ and ≥ are defined). We use ∞ to denote infinity:

Definition 2.2: Addition, scalar multiplication, comparison, and **size** of multi-sets are defined in the following way, for all $m, m_1, m_2 \in S_{MS}$ and all $n \in \mathbb{N}$:

(i) $m_1 + m_2 = \sum_{s \in S} (m_1(s) + m_2(s))`s$ (addition).

(ii) $n * m = \sum_{s \in S} (n * m(s))`s$ (scalar multiplication).

(iii) $m_1 \neq m_2 = \exists s \in S: m_1(s) \neq m_2(s)$
$m_1 \leq m_2 = \forall s \in S: m_1(s) \leq m_2(s)$ (comparison; ≥ and = are defined analogously to ≤).

(iv) $|m| = \sum_{s \in S} m(s)$ (size).

When $|m| = \infty$ we say that m is **infinite**. Otherwise m is **finite**. When $m_1 \leq m_2$ we also define **subtraction**:

(v) $m_2 - m_1 = \sum_{s \in S} (m_2(s) - m_1(s))`s$ (subtraction).

The reader is invited to check that the formal definitions of the operations in Def. 2.2 are consistent with the informal explanations given in Fig. 2.1. The multi-set operations have a number of nice properties, some of which are listed below. We use $\mathbb{N}_+$ to denote the set of all positive integers. Moreover, we define $n + \infty = \infty + n = \infty$ for all $n \in \mathbb{N} \cup \{\infty\}$, while $n * \infty = \infty * n = \infty$ for all $n \in \mathbb{N}_+ \cup \{\infty\}$ and $0 * \infty = \infty * 0 = 0$.

Proposition 2.3: The following equations are satisfied for all multi-sets $m, m_1, m_2, m_3 \in S_{MS}$ and all non-negative integers $n, n_1, n_2 \in \mathbb{N}$:

(i) $m_1 + m_2 = m_2 + m_1$ (+ is commutative).
(ii) $m_1 + (m_2 + m_3) = (m_1 + m_2) + m_3$ (+ is associative).
(iii) $m + \emptyset = m$ (∅ is zero element).
(iv) $1 * m = m$ (1 is one element).
(v) $0 * m = \emptyset$.
(vi) $n * (m_1 + m_2) = (n * m_1) + (n * m_2)$.
(vii) $(n_1 + n_2) * m = (n_1 * m) + (n_2 * m)$.
(viii) $n_1 * (n_2 * m) = (n_1 * n_2) * m$.
(ix) $|m_1 + m_2| = |m_1| + |m_2|$.
(x) $|n * m| = n * |m|$.

Proof: The proof is straightforward, and hence it is omitted. □

In terms of algebra, Prop. 2.3 (i)–(iii) imply that $(S_{MS}, +)$ is a **commutative monoid**. Finally, we define sums over multi-sets:

Definition 2.4: Let a finite multi-set $m \in S_{MS}$, a commutative monoid (R,+) and a function $F \in [S \to R]$ be given. We then use:

$$\sum_{s \in m} F(s)$$

to denote the **sum** which has an addend for each element of m. When an element of S appears several times in the multi-set m, there is an addend for each of these appearances.

As an example of this notation, each multi-set m satisfies:

$$m = \sum_{s \in m} s.$$

2.2 Structure of Non-Hierarchical CP-nets

In this section we define non-hierarchical CP-nets as many-tuples. However, it should be understood that the only purpose of this is to give a mathematically sound and unambiguous definition of CP-nets and their semantics. Any particular net – created by a modeller – will always be specified in terms of a CPN diagram, i.e., a diagram similar to Figs. 1.7 and 1.13. Our approach is analogous to the definition of directed graphs and non-deterministic finite automata. Formally, they are defined as pairs G = (V, A) and as 5-tuples $M = (Q, \Sigma, \delta, q_0, F)$, respectively, but usually they are represented by drawings containing sets of nodes, arcs and inscriptions.

To give the abstract definition of CP-nets it is not necessary to fix the concrete syntax in which the modeller writes the net expressions, and thus we shall only assume that such a syntax exists together with a well-defined semantics – making it possible in an unambiguous way to talk about:

- The *elements of a type*, T. The set of all elements in T is denoted by the type name T itself.
- The *type of a variable*, v – denoted by Type(v).
- The *type of an expression*, expr – denoted by Type(expr).
- The *set of variables in an expression*, expr – denoted by Var(expr).
- A *binding of a set of variables*, V – associating with each variable $v \in V$ an element $b(v) \in Type(v)$.
- The *value obtained by evaluating an expression*, expr, *in a binding*, b – denoted by expr<b>. Var(expr) is required to be a subset of the variables of b, and the evaluation is performed by substituting for each variable $v \in Var(expr)$ the value $b(v) \in Type(v)$ determined by the binding.

An expression *without* variables is said to be a **closed** expression. It can be evaluated in all bindings, and all evaluations give the same value – which we often shall denote by the expression itself. This means that we simply write "expr" instead of the more pedantic "expr<b>".

Now we are ready to define non-hierarchical CP-nets. We use $\mathbb{B}$ to denote the boolean type (containing the elements {false, true}, and having the standard operations from propositional logic), and when Vars is a set of variables we use Type(Vars) to denote the set of types $\{Type(v) \mid v \in Vars\}$. Some motivations and explanations of the individual parts of the definition are given immediately below the definition, and it is recommended that these are read in parallel with the definition. A similar remark applies to many other definitions in this book – in particular those which introduce a many-tuple.

Definition 2.5: A **non-hierarchical CP-net** is a tuple CPN = $(\Sigma, P, T, A, N, C, G, E, I)$ satisfying the requirements below:

(i) Σ is a finite set of non-empty types, called **colour sets**.
(ii) P is a finite set of **places**.
(iii) T is a finite set of **transitions**.
(iv) A is a finite set of **arcs** such that:
- $P \cap T = P \cap A = T \cap A = \emptyset$.

(v) N is a **node** function. It is defined from A into $P \times T \cup T \times P$.
(vi) C is a **colour** function. It is defined from P into Σ.
(vii) G is a **guard** function. It is defined from T into expressions such that:
- $\forall t \in T: [Type(G(t)) = \mathbb{B} \land Type(Var(G(t))) \subseteq \Sigma]$.

(viii) E is an **arc expression** function. It is defined from A into expressions such that:
- $\forall a \in A: [Type(E(a)) = C(p(a))_{MS} \land Type(Var(E(a))) \subseteq \Sigma]$

where p(a) is the place of N(a).
(ix) I is an **initialization** function. It is defined from P into closed expressions such that:
- $\forall p \in P: [Type(I(p)) = C(p)_{MS}]$.

(i) The set of **colour sets** determines the types, operations and functions that can be used in the net inscriptions (i.e., arc expressions, guards, initialization expressions, colour sets, etc.). If desired, the colour sets (and the corresponding operations and functions) can be defined by means of a many-sorted sigma algebra (as in the theory of abstract data types). We assume that each colour set has at least one element.

(ii) + (iii) + (iv) The **places, transitions** and **arcs** are described by three sets P, T and A which are required to be finite and pairwise disjoint. In contrast to classical Petri nets, we allow the net structure to be empty (i.e., $P \cup T \cup A = \emptyset$). The reason is pragmatic. It allows the user to define and syntax-check a set of colour sets without having to invent a dummy net structure. We require the sets of places, transitions and arcs to be finite. This means that we avoid a number of technical problems such as the possibility of having an infinite number of arcs between two nodes.

(v) The **node** function maps each arc into a pair where the first element is the source node and the second the destination node. The two nodes have to be of different kind (i.e., one must be a place while the other is a transition). In contrast to classical Petri nets, we allow a CP-net to have several arcs between the same ordered pair of nodes (and thus we define A as a separate set and not as a subset of $P \times T \cup T \times P$). The reason is again pragmatic. We often have nets where each occurring binding element moves exactly one token along each of the surrounding arcs, and it is then awkward to be forced to violate this convention in the cases where a binding element removes/adds two or more tokens to/from the same place. It is of course easy to combine such multiple arcs into a single arc by adding the corresponding arc expressions (that is always possible because they all have a common multi-set type). We also allow nodes to be isolated. Again the reason is pragmatic. When we build computer tools for CP-nets we want to be able to check whether a diagram is a CP-net (i.e., fulfils the definition above). However, there is no conceptual difference between an isolated node and a node where all the arc expressions of the surrounding arcs always evaluate to the empty multi-set (and the latter is difficult to detect in general, since arc expressions may be arbitrarily complex). It is of course easy to exclude such degenerate nets when this is convenient for theoretical purposes.

(vi) The **colour** function C maps each place, p, to a colour set C(p). Intuitively, this means that each token on p must have a token colour that belongs to the type C(p).

(vii) The **guard** function G maps each transition, t, to an expression of type boolean, i.e., a predicate. Moreover, all variables in G(t) must have types that belong to Σ. We shall also allow a guard to be a *list* of boolean expressions [$Bexpr_1$, $Bexpr_2$, ..., $Bexpr_n$], and this is a shorthand for the boolean expression $Bexpr_1 \wedge Bexpr_2 \wedge \ldots \wedge Bexpr_n$. Intuitively, this means that the binding must fulfil each of the boolean expressions in the list. We shall allow a guard expression to be missing, and consider this to be a shorthand for the closed expression true.

(viii) The **arc expression** function E maps each arc, a, into an expression which must be of type $C(p(a))_{MS}$. This means that each evaluation of the arc expression must yield a multi-set over the colour set that is attached to the corresponding place. We shall also allow a CPN diagram to have an arc expression expr of type C(p(a)), and consider this to be a shorthand for 1\`(expr). Intuitively, this means that the arc expression evaluates to a colour in C(p(a)) which we then consider to be a multi-set with only one element. Finally, we shall allow an arc expression to be missing, and consider this to be a shorthand for empty.

(ix) The **initialization** function I maps each place, p, into a closed expression which must be of type $C(p)_{MS}$, i.e., a multi-set over C(p). Analogously to (viii), we shall also allow, as a shorthand, an initial expression to be of type C(p) or to be missing.

As mentioned in the preface, the "modern version" of CP-nets (presented in this book) uses the expression representation (defined above) not only when a system is being described, but also when it is being analysed. It is only during invariants analysis that it may be adequate/necessary to translate the expression representation into a function representation.

In addition to the concepts introduced in Def. 2.5, we use $X = P \cup T$ to denote the set of all **nodes**. Moreover, we define a number of functions describing the relationship between neighbouring elements of the net structure. Each function name indicates the range of the function. As an example, p maps into places, while A maps into sets of arcs (a capital letter indicates that the elements of the range are sets). We sometimes use the same name for several functions/sets, but from the argument(s) it will always be clear which one we deal with, e.g., the set A, the function $A \in [X \rightarrow A_S]$ or the function $A \in [(P \times T \cup T \times P) \rightarrow A_S]$:

- $p \in [A \rightarrow P]$ maps each arc, a, to the **place** of N(a), i.e., that component of N(a) which is a place.
- $t \in [A \rightarrow T]$ maps each arc, a, to the **transition** of N(a), i.e., that component of N(a) which is a transition.
- $s \in [A \rightarrow X]$ maps each arc, a, to the **source** of a, i.e., the first component of N(a).
- $d \in [A \rightarrow X]$ maps each arc, a, to the **destination** of a, i.e., the second component of N(a).
- $A \in [(P \times T \cup T \times P) \rightarrow A_S]$ maps each ordered pair of nodes (x_1, x_2) to the set of its **connecting arcs**, i.e., the arcs that have the first node as source and the second as destination:
  $A(x_1, x_2) = \{a \in A \mid N(a) = (x_1, x_2)\}$.
- $A \in [X \rightarrow A_S]$ maps each node x to the set of its **surrounding arcs**, i.e., the arcs that have x as source or destination:
  $A(x) = \{a \in A \mid \exists x' \in X: [N(a) = (x, x') \vee N(a) = (x', x)]\}$.
- $In \in [X \rightarrow X_S]$ maps each node x to the set of its **input nodes**, i.e., the nodes that are connected to x by an input arc:
  $In(x) = \{x' \in X \mid \exists a \in A: N(a) = (x', x)\}$.
- $Out \in [X \rightarrow X_S]$ maps each node x to the set of its **output nodes**, i.e., the nodes that are connected to x by an output arc:
  $Out(x) = \{x' \in X \mid \exists a \in A: N(a) = (x, x')\}$.
- $X \in [X \rightarrow X_S]$ maps each node x to the set of its **surrounding nodes**, i.e., the nodes that are connected to x by an arc:
  $X(x) = \{x' \in X \mid \exists a \in A: [N(a) = (x, x') \vee N(a) = (x', x)]\}$.

All the functions above can be extended, in the usual way, to take sets as input (then they all return sets and thus all the function names are written with a capital letter). We shall often use •x instead of In(x), and x• instead of Out(x) – and this will also be done when x is a set of nodes.

To illustrate our formal definition of non-hierarchical CP-nets, Fig. 2.2 shows how the net in Fig. 1.7 can be represented as a many-tuple (omitting the detailed specification of the four colour sets U, I, P and E). Notice that some of the names are used for multiple purposes. As an example, E is the name of a colour set, the name of a place, and the name of the arc expression function. From the contents it will always be clear which of these we refer to. This will be the only time in this book that we present a particular non-hierarchical CP-net as a tuple. Hopefully, Fig. 2.2 demonstrates that, in practice, it would be intolerable for humans to work with non-hierarchical CP-nets – without having a graphical representation of them.

(i) $\Sigma = \{U, I, P, E\}$.

(ii) $P = \{A, B, C, D, E, R, S, T\}$.

(iii) $T = \{T1, T2, T3, T4, T5\}$.

(iv) $A = \{$AtoT1, T1toB, BtoT2, T2toC, CtoT3, T3toD, DtoT4, T4toE, EtoT5, T5toA, T5toB, RtoT1, StoT1, StoT2, TtoT3, TtoT4, T3toR, T5toS, T5toT$\}$.

(v) N(a) = (SOURCE,DEST) if a is in the form SOURCEtoDEST.

(vi)
$$C(p) = \begin{cases} P & \text{if } p \in \{A, B, C, D, E\} \\ E & \text{otherwise.} \end{cases}$$

(vii)
$$G(t) = \begin{cases} x{=}q & \text{if } t{=}T1 \\ \text{true} & \text{otherwise.} \end{cases}$$

(viii)
$$E(a) = \begin{cases} \text{e} & \text{if } a \in \{\text{RtoT1, StoT1, TtoT4}\} \\ 2\text{`e} & \text{if } a = \text{T5toS} \\ \text{case x of p=>2`e | q=>1`e} & \text{if } a \in \{\text{StoT2, T5toT}\} \\ \text{if x=q then 1`e else empty} & \text{if } a = \text{T3toR} \\ \text{if x=p then 1`e else empty} & \text{if } a = \text{TtoT3} \\ \text{if x=q then 1`(q,i+1) else empty} & \text{if } a = \text{T5toA} \\ \text{if x=p then 1`(p,i+1) else empty} & \text{if } a = \text{T5toB} \\ \text{(x,i)} & \text{otherwise.} \end{cases}$$

(ix)
$$I(p) = \begin{cases} 3\text{`(q,0)} & \text{if } p{=}A \\ 2\text{`(p,0)} & \text{if } p{=}B \\ 1\text{`e} & \text{if } p{=}R \\ 3\text{`e} & \text{if } p{=}S \\ 2\text{`e} & \text{if } p{=}T \\ \emptyset & \text{otherwise.} \end{cases}$$

Fig. 2.2. The CP-net from Fig. 1.7 represented as a many-tuple

2.3 Behaviour of Non-Hierarchical CP-nets

Having defined the structure of CP-nets we are now ready to consider their behaviour – but first we introduce the following notation:

- $\forall t \in T$: $Var(t) = \{v \mid v \in Var(G(t)) \vee \exists a \in A(t)\colon v \in Var(E(a))\}$.
- $\forall (x_1,x_2) \in (P \times T \cup T \times P)$: $E(x_1,x_2) = \sum_{a \in A(x_1,x_2)} E(a)$.

Var(t) is called the set of **variables** of t while $E(x_1,x_2)$ is called the **expression** of (x_1,x_2). The summation indicates addition of expressions (and it is well-defined because all the participating expressions have a common multi-set type). From the argument(s) it will always be clear whether we deal with the function $E \in [A \rightarrow Expr]$ or the function $E \in [(P \times T \cup T \times P) \rightarrow Expr]$. Notice that $A(x_1,x_2) = \emptyset$ implies that $E(x_1,x_2) = \emptyset$ (where the latter $\emptyset$ denotes the closed expression which evaluates to the empty multi-set). Next we define what we mean by a transition binding. Intuitively, a binding of a transition t is a substitution that replaces each variable of t with a colour. It is required that each colour is of the correct type and that the guard evaluates to true. As defined in Sect. 2.2, G(t)<b> denotes the evaluation of the guard expression G(t) in the binding b:

Definition 2.6: A **binding** of a transition t is a function b defined on Var(t), such that:

(i) $\forall v \in Var(t)$: $b(v) \in Type(v)$.

(ii) G(t)<b>.

By B(t) we denote the set of all bindings for t.

As shown in Sect. 1.2 we usually write bindings in the form $<v_1=c_1,v_2=c_2,\ldots,v_n=c_n>$ where $Var(t) = \{v_1,v_2,\ldots,v_n\}$. The order of the variables has no importance, and this means, e.g., that the binding <x=q,i=0> is the same as <i=0,x=q>. Next we define token elements, binding elements, markings and steps:

Definition 2.7: A **token element** is a pair (p,c) where $p \in P$ and $c \in C(p)$, while a **binding element** is a pair (t,b) where $t \in T$ and $b \in B(t)$. The set of all token elements is denoted by TE while the set of all binding elements is denoted by BE.

A **marking** is a multi-set over TE while a **step** is a *non-empty* and *finite* multi-set over BE. The **initial marking** M_0 is the marking which is obtained by evaluating the initialization expressions:

$$\forall (p,c) \in TE: M_0(p,c) = (I(p))(c).$$

The sets of all markings and steps are denoted by $\mathbb{M}$ and $\mathbb{Y}$, respectively.

Each marking $M \in TE_{MS}$ determines a unique function M^* defined on P such that $M^*(p) \in C(p)_{MS}$:

$$\forall p \in P \; \forall c \in C(p): (M^*(p))(c) = M(p,c).$$

On the other hand, each function M^*, defined on P such that $M^*(p) \in C(p)_{MS}$ for all $p \in P$, determines a unique marking M:

$$\forall (p,c) \in TE: M(p,c) = (M^*(p))(c).$$

Thus we shall often represent markings as functions defined on P (and we shall use the same name for the function and the multi-set representation of a marking). Analogously, there is a unique correspondence between a step Y and a function Y* defined on T such that $Y^*(t) \in B(t)_{MS}$ is finite for all $t \in T$ and non-empty for at least one $t \in T$. Hence we often represent steps as functions defined on T.

As an example, we can represent the initial marking of the CP-net in Fig. 1.7 either as a function:

$$M_0(p) = \begin{cases} 3`(q,0) & \text{if } p=A \\ 2`(p,0) & \text{if } p=B \\ 1`e & \text{if } p=R \\ 3`e & \text{if } p=S \\ 2`e & \text{if } p=T \\ \emptyset & \text{otherwise,} \end{cases}$$

or as a multi-set:

$$M_0 = 3`(A,(q,0)) + 2`(B,(p,0)) + 1`(R,e) + 3`(S,e) + 2`(T,e).$$

Now we are ready to give the formal definition of enabling (in which we represent steps by multi-sets and functions, while we represent markings by functions only). The expression evaluation $E(p,t)\langle b\rangle$ yields the multi-set of token colours, which are removed from p when t occurs with the binding b. By taking the sum over all binding elements $(t,b) \in Y$ we get all the tokens that are removed from p when Y occurs. This multi-set is required to be less than or equal to the marking of p. It means that each binding element $(t,b) \in Y$ must be able to get the tokens specified by $E(p,t)\langle b\rangle$, without having to share these tokens with other binding elements of Y. It should be remembered that all bindings of a step, according to Def. 2.6, automatically satisfy the corresponding guards. Moreover, it should be noted that the summations in Defs. 2.8 and 2.9 are summations over multi-sets (cf. Def. 2.4).

Definition 2.8: A step Y is **enabled** in a marking M iff the following property is satisfied:

$$\forall p \in P: \sum_{(t,b) \in Y} E(p,t)\langle b\rangle \leq M(p).$$

Let the step Y be enabled in the marking M. When $(t,b) \in Y$, we say that t is **enabled** in M for the binding b. We also say that (t,b) is enabled in M, and so is t. When $(t_1,b_1),(t_2,b_2) \in Y$ and $(t_1,b_1) \neq (t_2,b_2)$ we say that (t_1,b_1) and (t_2,b_2) are **concurrently enabled**, and so are t_1 and t_2. When $|Y(t)| \geq 2$ we say that t is concurrently enabled with itself. When $Y(t,b) \geq 2$ we say that (t,b) is concurrently enabled with itself.

When a step is enabled it may occur, and this means that tokens are removed from the input places and added to the output places of the occurring transitions. The number and colours of the tokens are determined by the arc expressions, evaluated for the occurring bindings:

Definition 2.9: When a step Y is enabled in a marking M_1 it may **occur,** changing the marking M_1 to another marking M_2, defined by:

$$\forall p \in P: M_2(p) = \Big(M_1(p) - \sum_{(t,b)\in Y} E(p,t)\langle b\rangle\Big) + \sum_{(t,b)\in Y} E(t,p)\langle b\rangle.$$

The first sum is called the **removed** tokens while the second is called the **added** tokens. Moreover, we say that M_2 is **directly reachable** from M_1 by the occurrence of the step Y, which we also denote: $M_1\,[Y\rangle\, M_2$.

The occurrence of a step is an indivisible event. Although the formula above requires the subtraction to be performed before the addition we do *not* recognize the existence of an intermediate marking, where the removed tokens have been removed while the added tokens have not yet been added. It should also be noted that a step does *not* need to be maximal. When a number of binding elements are concurrently enabled, it is possible to have an occurring step which only contains some of them.

To illustrate Defs. 2.8 and 2.9, let us now consider the data base system from Sect. 1.3 (Fig. 1.13). We use SM, RM, SA, and RA as abbreviations for the transition names and we denote a step by listing those pairs (t,Y(t)) where Y(t) differs from the empty multi-set. In the initial marking M_0 (which we also shall call M_1) the following four steps are enabled:

Y_{1a} = (SM, 1`<s=d_1>)
Y_{1b} = (SM, 1`<s=d_2>)
Y_{1c} = (SM, 1`<s=d_3>)
Y_{1d} = (SM, 1`<s=d_4>).

Each of these steps contains only a single binding element. No other steps are enabled in M_1. If Y_{1a} occurs in M_1, we get a marking M_2 in which the following step is enabled:

Y_{2a} = (RM, 1`<s=$d_1$,r=$d_2$> + 1`<s=d_1,r=d_3> + 1`<s=d_1,r=d_4>).

In this step RM is concurrently enabled with itself. When a step is enabled in a given marking, it is trivial to verify that each smaller step is also enabled. This means that, e.g., the following step also is enabled in M_2:

Y_{2b} = (RM, 1`<s=$d_1$,r=$d_2$> + 1`<s=d_1,r=d_4>).

If Y_{2b} occurs in M_2, we get a marking M_3 in which the following step is enabled:

Y_{3a} = (RM, 1`<s=d_1,r=d_3>),
(SA, 1`<s=$d_1$,r=$d_2$> + 1`<s=d_1,r=d_4>).

This means that, e.g., the following step also is enabled in M_3:

Y_{3b} = (RM, 1`<s=d_1,r=d_3>),
(SA, 1`<s=d_1,r=d_4>).

If Y_{3b} occurs in M_3, we get a marking M_4 in which the following step is enabled:

Y_{4a} = (SA, 1`<s=$d_1$,r=$d_2$> + 1`<s=d_1,r=d_3>).

If Y_{4a} occurs in M_4, we get a marking M_5 in which the following is the only enabled step:

$$Y_{5a} = (RA, 1`<s=d_1>).$$

If Y_{5a} occurs in M_5, we get a marking M_6 which is identical to the initial marking. In the example above we have considered sequences of reachable markings and occurring steps. This can be formalized as shown below. We use 1..n to denote the set of all integers i that satisfy $1 \leq i \leq n$.

Definition 2.10: A **finite occurrence sequence** is a sequence of markings and steps:

$$M_1 [Y_1\rangle M_2 [Y_2\rangle M_3 \ldots M_n [Y_n\rangle M_{n+1}$$

such that $n \in \mathbb{N}$, and $M_i [Y_i\rangle M_{i+1}$ for all $i \in 1..n$. The marking M_1 is called the **start marking** of the occurrence sequence, while the marking M_{n+1} is called the **end marking**. The non-negative integer n is called the **number of steps** in the occurrence sequence, or the **length** of it.

Analogously, an **infinite occurrence sequence** is a sequence of markings and steps:

$$M_1 [Y_1\rangle M_2 [Y_2\rangle M_3 \ldots$$

such that $M_i [Y_i\rangle M_{i+1}$ for all $i \in \mathbb{N}_+$. The marking M_1 is called the **start marking** of the occurrence sequence, which is said to have **infinite length**.

The set of all finite occurrence sequences is denoted by OSF, while the set of all infinite occurrence sequences is denoted by OSI. Moreover, we use $OS = OSF \cup OSI$ to denote the set of all occurrence sequences.

The start marking of an occurrence sequence will often, but not always, be identical to the initial marking of the CP-net. Above we have seen that the data base system has the following finite occurrence sequence, of length 5:

$$M_1 [Y_{1a}\rangle M_2 [Y_{2b}\rangle M_3 [Y_{3b}\rangle M_4 [Y_{4a}\rangle M_5 [Y_{5a}\rangle M_6.$$

We allow the user to omit some parts of an occurrence sequence and, e.g., write:

$$M_1 [Y_{1a}\ Y_{2b}\ Y_{3b}\ Y_{4a}\ Y_{5a}\rangle M_6 \quad \text{or} \quad M_1 [Y_{1a}\ Y_{2b}\ Y_{3b}\ Y_{4a}\ Y_{5a}\rangle.$$

Both of these abbreviated forms are equivalent to the original form – in the sense that the missing markings can be recalculated from the abbreviated form. As a particular case of the second abbreviation, we shall use the following notation to denote that the binding element (SM,<s=d1>) and the transition SM is enabled in M_1:

$$M_1 [SM,<s=d_1>\rangle \quad \text{and} \quad M_1 [SM\rangle.$$

It is also possible to abbreviate finite occurrence sequences by omitting the steps, e.g.:

$M_1 [1\rangle M_2 [1\rangle M_3 [1\rangle M_4 [1\rangle M_5 [1\rangle M_6$ or

$M_1 [3\rangle M_4 [2\rangle M_6$ or $M_1 [5\rangle M_6$ or $M_1 [*\rangle M_6$.

From these abbreviated forms, it is *not* always possible to recalculate the original form, e.g., because two enabled steps may have the same effect on the marking. Infinite occurrence sequences can be abbreviated in similar ways, e.g.:

$M_1 [Y_1 Y_2 Y_3 \ldots\rangle$ or $M_1 [1\rangle M_2 [1\rangle M_3 [1\rangle \ldots$

Definition 2.11: A marking M" is **reachable** from a marking M' iff there exists a finite occurrence sequence having M' as start marking and M" as end marking, i.e., iff for some $n \in \mathbb{N}$ there exists a sequence of steps $Y_1 Y_2 \ldots Y_n$ such that:

$M' [Y_1 Y_2 \ldots Y_n\rangle M''$.

We then also say that M" is reachable from M' in n steps. As a shorthand, we say that a marking is reachable iff it is reachable from M_0. The *set* of markings which are reachable from M' is denoted by $[M'\rangle$.

Notice that we allow occurrence sequences with length zero. This means that $M \in [M\rangle$ for all markings M. Moreover, it means that we must distinguish between the following two statements (where ε denotes the empty sequence while * denotes an unknown sequence):

$M' [*\rangle M''$ and $M' [\varepsilon\rangle M''$.

The first of these statements only tells us that $M'' \in [M'\rangle$, while the second tells us that $M'' = M'$.

In Vol. 2 of this work we shall see that the concepts of enabling, occurrence, and reachability can also be expressed by means of linear algebra. To do this we represent a CP-net as a matrix, where each individual matrix element is a linear function mapping multi-sets of bindings to multi-sets of token colours.

2.4 Equivalent Place/Transition Nets

This section is theoretical and it can be skipped by readers who are primarily interested in the practical application of CP-nets. The section defines Place/Transition Nets (PT-nets) and we investigate the formal relationship between CP-nets and PT-nets. For each non-hierarchical CP-net we show how to construct an equivalent PT-net, i.e., a PT-net which has exactly the same behaviour as the CP-net. The existence of an equivalent PT-net is extremely useful, because it tells us how to generalize the basic concepts and analysis methods of PT-nets to non-hierarchical CP-nets. We simply define these concepts in such a way that a CP-net has a given property iff the equivalent PT-net has the corresponding property. It is important to understand that we never make the translation for a particular non-hierarchical CP-net. When we describe a system we use non-hierarchical CP-nets directly, without constructing the equivalent PT-net. Analogously, we analyse a non-hierarchical CP-net directly, without constructing the equivalent PT-net. First we define PT-nets:

Definition 2.12: A **Place/Transition Net** is a tuple PTN = (P, T, A, E, I) satisfying the requirements below:

(i) P is a set of **places**.
(ii) T is a set of **transitions** such that:
- $P \cap T = \emptyset$.

(iii) $A \subseteq P \times T \cup T \times P$ is a set of **arcs**.
(iv) $E \in [A \to \mathbb{N}_+]$ is an **arc expression** function.
(v) $I \in [P \to \mathbb{N}]$ is an **initialization** function.

(i) + (ii) The **places** and **transitions** are described by two sets P and T which are required to be disjoint, but allowed to be infinite. In contrast to the classical definition of PT-nets, we allow the set of nodes $X = P \cup T$ to be empty.

(iii) The **arcs** are described as a subset of $P \times T \cup T \times P$, and this indicates that an arc either goes from a place to a transition or from a transition to a place. In contrast to the classical definition of PT-nets, we allow places and transitions to be isolated (i.e., to be without surrounding arcs). It is of course easy to remove such isolated nodes when this is convenient for theoretical purposes.

(iv) + (v) The **arc expression** function E maps each arc to a positive integer, while the **initialization** function maps each place to a non-negative integer.

In contrast to the classical definition of PT-nets we do not have a capacity function. However, it is a well-known fact that each PT-net with capacities can be translated into an equivalent PT-net without capacities.

Having defined the structure of PT-nets we can now define their behaviour. We define p(a), t(a), s(a), d(a), $A(x_1,x_2)$, A(x), In(x), Out(x), X(x) and $E(x_1,x_2)$ in a way which is analogous to what we did for non-hierarchical CP-nets. These definitions are straightforward and thus they are omitted.

Definition 2.13: A **marking** is a multi-set over P while a **step** is a *non-empty* and *finite* multi-set over T. The **initial marking** M_0 is the marking which is obtained from the initialization expressions:

$$\forall p \in P: M_0(p) = I(p).$$

The sets of all markings and steps are denoted by $\mathbb{M}$ and $\mathbb{Y}$, respectively.

It should be recalled that multi-sets are functions into $\mathbb{N}$, and this means that a marking is a function $M \in [P \to \mathbb{N}]$ while a step is a function $Y \in [T \to \mathbb{N}]$. Next we define enabling and occurrence. Notice that the summation is over a multi-set Y in which there may be several appearances of t (cf. Def. 2.4).

Definition 2.14: A step Y is **enabled** in a marking M iff the following property is satisfied:

$$\forall p \in P: \sum_{t \in Y} E(p,t) \leq M(p).$$

Let the step Y be enabled in the marking M. When $t \in Y$, we say that t is **enabled** in M. When $t_1, t_2 \in Y$ and $t_1 \neq t_2$ we say that t_1 and t_2 are **concurrently enabled**. When $Y(t) \geq 2$ we say that t is concurrently enabled with itself.

When a step Y is enabled in a marking M_1 it may **occur**, changing the marking M_1 to another marking M_2, defined by:

$$\forall p \in P: M_2(p) = \Big(M_1(p) - \sum_{t \in Y} E(p,t)\Big) + \sum_{t \in Y} E(t,p).$$

We define **removed/added tokens, direct reachability, occurrence sequences** and **reachability** analogously to the corresponding concepts for non-hierarchical CP-nets.

Now we can define how to translate a non-hierarchical CP-net into a behaviourally equivalent PT-net. It should of course be verified that the defined tuple really is a PT-net, i.e., that it satisfies the constraints in Def. 2.12. However, this verification is straightforward and thus it is omitted.

Definition 2.15: Let a non-hierarchical CP-net CPN = $(\Sigma, P, T, A, N, C, G, E, I)$ be given. Then we define the **equivalent PT-net** to be PTN* = $(P^*, T^*, A^*, E^*, I^*)$ where:

(i) $P^* = TE$.

(ii) $T^* = BE$.

(iii) $A^* = \{((p,c),(t,b)) \in P^* \times T^* \mid (E(p,t)<b>)(c) \neq 0\} \cup \{((t,b),(p,c)) \in T^* \times P^* \mid (E(t,p)<b>)(c) \neq 0\}$.

(iv) $\forall ((p,c),(t,b)) \in A^* \cap (P^* \times T^*): E^*((p,c),(t,b)) = (E(p,t)<b>)(c)$
$\forall ((t,b),(p,c)) \in A^* \cap (T^* \times P^*): E^*((t,b),(p,c)) = (E(t,p)<b>)(c)$.

(v) $\forall (p,c) \in P^*: I^*(p,c) = (I(p))(c)$.

(i) The PT-net has a place for each token element in the CP-net. This means that each CP-net place p is split into as many PT-net places as there are token colours in C(p). In this way we are, after the translation, still able to distinguish between tokens which had different colours. The tokens have "lost" their colours – but now they are positioned at different places.

(ii) The PT-net has a transition for each binding element in the CP-net. This means that each CP-net transition t is split into as many PT-net transitions as there are bindings in B(t). It should be remembered that all elements of B(t), according to Def. 2.6, automatically satisfy the guard G(t).

(iii) There is an arc from a PT-net place (p,c) to a PT-net transition (t,b) iff $(E(p,t)<b>)(c) \neq 0$, i.e., iff an occurrence of t with the binding b removes at least one c-token from p. Analogously, there is an arc from a PT-net transition (t,b) to a PT-net place (p,c) iff $(E(t,p)<b>)(c) \neq 0$, i.e., iff an occurrence of t with the binding b adds at least one c-token to p.

(iv) The arc expression $E^*((p,c),(t,b))$ of an arc from a PT-net place (p,c) to a PT-net transition (t,b) is defined as $(E(p,t)<b>)(c)$, i.e., the number of c-tokens which an occurrence of t with the binding b removes from p. Analogously, the arc expression $E^*((t,b),(p,c))$ of an arc from a PT-net transition (t,b) to a PT-net place (p,c) is defined as $(E(t,p)<b>)(c)$, i.e., the number of c-tokens which an occurrence of t with the binding b adds to p.

(v) The initialization expression $I^*(p,c)$ of a PT-net place (p,c) is defined as $(I(p))(c)$, i.e., the number of c-tokens in I(p).

The following theorem shows that each non-hierarchical CP-net has exactly the same sets of markings, steps and occurrence sequences as the equivalent PT-net, and thus the two nets are behaviourally equivalent. All concepts with a star refer to PTN*, while those without refer to CPN:

Theorem 2.16: Let CPN be a non-hierarchical CP-net and let PTN* be the equivalent PT-net. Then we have the following properties:

(i) $\mathbb{M} = \mathbb{M}^* \wedge M_0 = M_0^*$.

(ii) $\mathbb{Y} = \mathbb{Y}^*$.

(iii) $\forall M_1,M_2 \in \mathbb{M}\ \forall Y \in \mathbb{Y}$: $M_1 [Y\rangle_{CPN} M_2 \Leftrightarrow M_1 [Y\rangle_{PTN^*} M_2$.

Proof: The proof is a simple consequence of our earlier definitions.

Property (i): From Def 2.7 and Def 2.13 we have that $\mathbb{M} = TE_{MS}$ while $\mathbb{M}^* = P^*_{MS}$. Thus it is sufficient to prove that $P^* = TE$, but this follows from Def. 2.15 (i).

Next let us prove that the two initial markings are identical. From Def. 2.7 we have:

(*) $\forall (p,c) \in TE$: $M_0(p,c) = (I(p))(c)$.

From Def. 2.13 we have:

$$\forall p^* \in P^*: M_0^*(p^*) = I^*(p^*),$$

which by Def. 2.15 (i) is equivalent to:

$$\forall (p,c) \in TE: M_0^*(p,c) = I^*(p,c),$$

which by Def. 2.15 (v) is equivalent to:

$$\forall (p,c) \in TE: M_0^*(p,c) = (I(p))(c),$$

which has the same form as (*).

Property (ii): From Def 2.7 and Def 2.13 we have that $\mathbb{Y}$ consists of all non-empty and finite multi-sets in BE_{MS}, while $\mathbb{Y}^*$ consists of all non-empty and finite multi-sets in T^*_{MS}. Thus it is sufficient to prove that $T^* = BE$, but this follows from Def. 2.15 (ii).

Property (iii): First we prove that the enabling rules coincide, i.e., that:

$$M_1\,[Y\rangle_{CPN} \Leftrightarrow M_1\,[Y\rangle_{PTN^*}.$$

From Def 2.8 it follows that $M_1\,[Y\rangle_{CPN}$ iff:

$$(**)\quad \forall p \in P: \sum_{(t,b)\in Y} E(p,t)\langle b\rangle \le M_1(p).$$

From Def. 2.14 it follows that $M_1\,[Y\rangle_{PTN^*}$ iff:

$$\forall p^* \in P^*: \sum_{t^*\in Y} E^*(p^*,t^*) \le M_1(p^*),$$

which by Def. 2.15 (i)+(ii) is equivalent to:

$$\forall (p,c) \in TE: \sum_{(t,b)\in Y} E^*((p,c),(t,b)) \le M_1(p,c),$$

which by Def. 2.15 (iii)+(iv), and the extension of E^* from A^* to $P^* \times T^* \cup T^* \times P^*$, is equivalent to:

$$\forall (p,c) \in TE: \sum_{(t,b)\in Y} (E(p,t)\langle b\rangle)(c) \le M_1(p,c),$$

which by Def. 2.2 (i) is equivalent to:

$$\forall p \in P\ \forall c \in C(p): \Big(\sum_{(t,b)\in Y} (E(p,t)\langle b\rangle) \Big)(c) \le (M_1(p))(c),$$

which by Def. 2.2 (iii) is equivalent to:

$$\forall p \in P: \sum_{(t,b)\in Y} (E(p,t)\langle b\rangle) \le M_1(p),$$

which is identical to (**).

Next we can prove that the occurrence rules coincide, i.e., that

$$M_1\,[Y\rangle_{CPN}\,M_2 \Leftrightarrow M_1\,[Y\rangle_{PTN^*}\,M_2.$$

However, this is done in a way which is totally analogous to the proof above, and thus we shall omit it. □

It is also possible to go in the other direction and construct a non-hierarchical CP-net from a given PT-net. When the net structure of the PT-net is finite, the equivalent CP-net can be constructed as follows:

- Keep the net structure (i.e., the places, transitions and arcs) unaltered.
- For each place, p, use a colour set E = {e} with a single colour.
- For each transition, use the closed expression true as guard.
- For each arc, a, use E(a)`e as arc expression.
- For each place, p, use I(p)`e as initialization expression.

The construction can, however, be done in many other ways. An arbitrary set of PT-net places can be folded into a single CP-net place. We simply introduce a colour for each of the PT-net places and in this way we are still able to distinguish between the tokens – although they now all reside on a single place.

Analogously, an arbitrary set of PT-net transitions can be folded into a single CP-net transition. We simply give the transition a single variable of a type which has an element for each of the PT-net transitions. In this way we get a binding element, i.e., a possible way of occurrence, for each of the PT-net transitions.

Now we can give the formal definition of the translation (Def. 2.17 below). A **partition** of a set S is a division of S into a number of components. The components are required to be non-empty subsets of S, and they must be pairwise disjoint and have S as their union. A partition is finite if it has a finite set of components. In (viii) we use $(v_{t^*}:t^*)$ to indicate that the expressions $E^*(p^*,t^*)$ and $E^*(t^*,p^*)$ have a variable v_{t^*} of type t^*. This notation is inspired by CPN ML, where one is always allowed to add the type to a variable. It should of course be verified that the defined tuple really is a non-hierarchical CP-net, i.e., that it satisfies the constraints in Def. 2.5. However, this verification is straightforward and thus it is omitted.

Definition 2.17: Let a PT-net PTN = (P, T, A, E, I) and two finite partitions PP and TT (of P and T) be given. Then we define the **equivalent non-hierarchical CP-net with respect to the partitions** to be CPN = (Σ*, P*, T*, A*, N*, C*, G*, E*, I*) where:

(i) $\Sigma^* = PP \cup TT$.
(ii) $P^* = PP$.
(iii) $T^* = TT$.
(iv) $A^* = \{(p^*,t^*) \in P^* \times T^* \mid \exists p \in p^* \; \exists t \in t^*: A(p,t) \neq \emptyset\} \cup \{(t^*,p^*) \in T^* \times P^* \mid \exists p \in p^* \; \exists t \in t^*: A(t,p) \neq \emptyset\}$.
(v) $\forall a^* = (x_1,x_2) \in A^*: N^*(x_1,x_2) = (x_1,x_2)$.
(vi) $\forall p^* \in P^*: C(p^*) = p^*$.
(vii) $\forall t^* \in T^*: G^*(t^*) = \text{true}$.
(viii) $\forall (p^*,t^*) \in A^* \cap (P^* \times T^*): E^*(p^*,t^*) = \sum_{p \in p^*} E(p,(v_{t^*}:t^*))`p$.
$\forall (t^*,p^*) \in A^* \cap (T^* \times P^*): E^*(t^*,p^*) = \sum_{p \in p^*} E((v_{t^*}:t^*),p)`p$.
(ix) $\forall p^* \in P^*: I^*(p^*) = \sum_{p \in p^*} I(p)`p$.

(i) For each component c, of PP or TT, we define a colour set – which we also denote by c. The colour set c has a colour for each element in the component c. The colour set does not have any implicitly declared operations or functions.

(ii) There is a CP-net place for each component of PP. Intuitively, this means that we fold all the corresponding PT-net places into a single CP-net place.

(iii) There is a CP-net transition for each component of TT. Intuitively, this means that we fold all the corresponding PT-net transitions into a single CP-net transition.

(iv) There is an arc from a CP-net place p* to a CP-net transition t* iff one of the PT-net places which was folded in p* had an arc to one of the PT-net transitions folded into t*. If the arc exists we denote it by (p*,t*) – which is unique, because it follows from the construction of A* that there never will be multiple arcs (i.e., arcs with identical source and identical destination). Analogously, there is an arc from a CP-net transition t* to a CP-net place p* iff one of the PT-net transitions which was folded in t* had an arc to one of the PT-net places folded into p*. If the arc exists we denote it by (t*,p*).

(v) The node function is trivial, because of the way in which we denote the arcs. It maps each arc $(x_1,x_2) \in A^*$ into (x_1,x_2), and this tells us that the source is x_1 while the destination is x_2.

(vi) Each place p* has attached the colour set p* that corresponds to the PP-component from which p* was constructed. This means that there are as many different colours for the CP-net place p* as there are elements in the PP-component p*, i.e., a colour for each PT-net place that was folded into p*. Intuitively, the colours make it possible still to distinguish between the tokens – although they now reside on a single place.

(vii) All guards are the closed expression "true".

(viii) Each arc from a CP-net place p* to a CP-net transition t* has an arc expression with a single variable v_{t^*} of type t*. This means that there are as many different bindings for the CP-net transition t* as there are elements in the TT-component t*, i.e., a binding for each PT-net transition that was folded into t*. The evaluation of E*(p*,t*) for a binding $<v_{t^*}=t>$ yields a CP-net token, of colour p, for each token which the PT-net transition t removes from the PT-net place p. Intuitively, this means that an occurrence of the CP-net transition t* with the binding $<v_{t^*}=t>$ removes the same tokens from the CP-net place p* as the occurrence of the PT-net transition t removes from all the PT-net places in the PP-component p*. However, now all these tokens are gathered on the CP-net place p* – and each of them has information attached saying which PT-net place the token corresponds to. Arcs from transitions to places have arc expressions which are defined analogously.

(ix) A place p* has an initial marking which has a token, of colour p, for each initial PT-net token on the PT-net place p. Intuitively, this means that we gather all the tokens of the PT-net places in the PP-component p* on the CP-net place p* – and attach information to each token to record which place the token came from.

The following theorem shows that each PT-net has exactly the same sets of markings, steps and occurrence sequences as the equivalent CP-net, and thus the two nets are behaviourally equivalent. All concepts with a star refer to CPN*, while those without refer to PTN:

Theorem 2.18: Let PTN be a PT-net and let CPN* be the equivalent non-hierarchical CP-net with respect to two arbitrary finite partitions. Then we have the following properties:

(i) $\mathbb{M} = \mathbb{M}^* \wedge M_0 = M_0^*$.

(ii) $\mathbb{Y} = \mathbb{Y}^*$.

(iii) $\forall M_1,M_2 \in \mathbb{M}\ \forall Y \in \mathbb{Y}: M_1 [Y\rangle_{PTN} M_2 \Leftrightarrow M_1 [Y\rangle_{CPN^*} M_2$.

Proof: The proof is analogous to the proof of Theorem 2.16, and hence is omitted. □

As simple corollaries of Theorem 2.18 we get the following extremes:

- Each PT-net has an equivalent CP-net with at most one place and one transition (use partitions with a single component).
- Each PT-net – with a finite number of places and transitions – has an equivalent CP-net with an identical net structure (use partitions where each component only has a single element).

It should be obvious that a good translation usually lies between these two extremes – and that it is found by choosing partitions which combine places which are "alike" and transitions which are "alike". However, there is very seldom a unique answer to what "alike" means. Some modellers prefer a small net structure and more complex arc expressions, while others prefer a larger net structure and simpler arc expressions. This is analogous to the use of procedures in a programming language. Some programmers define many different procedures which do nearly the same thing, while others combine all these operations in a single procedure.

Bibliographical Remarks

The notation and terminology defined in this chapter are to a large extent consistent with the basic Petri net notation and terminology defined in [7]. However, we use mnemonic names based on the English language (e.g., P and T for places and transitions), while those of [7] primarily are based on the original German terminology introduced by C.A. Petri.

The relationship between high-level nets and low-level Petri nets has been explained in several papers, e.g., [39], [41] and [51].

Exercises

Exercise 2.1.
Consider the resource allocation system from Sect. 1.2 (Fig. 1.7).

(a) Calculate all occurrence sequences which start in M_0 and have a length which is less than or equal to 3.

(b) Represent the occurrence sequences from (a) in a directed graph which has a node for each marking and an arc for each step. Such a graph is called an occurrence graph (to which we shall return in Chap. 5).

Exercise 2.2.
Consider the philosopher system from Exercise 1.6 (b).

(a) Represent the CPN diagram as a tuple (cf. Fig. 2.2). While you do this, convince yourself that all the requirements in Def. 2.5 are fulfilled.

(b) Calculate all occurrence sequences which start in M_0 and have a length which is less than or equal to 3.

(c) Represent the occurrence sequences from (b) in a directed graph which has a node for each marking and an arc for each step. Such a graph is called an occurrence graph (to which we shall return in Chap. 5).

Exercise 2.3.
Consider the distributed data base system from Sect. 1.3.

(a) Represent the CPN diagram in Fig. 1.13 as a tuple (cf. Fig. 2.2). While you do this, convince yourself that all the requirements in Def. 2.5 are fulfilled.

(b) Calculate the number of occurrence sequences which start in M_0 and have a length which is less than or equal to 3.

(c) Calculate the number of reachable markings.

Exercise 2.4.
Consider the model railway from Exercise 1.5 (b).

(a) Represent the CPN diagram as a tuple (cf. Fig. 2.2). While you do this, convince yourself that all the requirements in Def. 2.5 are fulfilled.

(b) Calculate all occurrence sequences which start in M_0 and have a length which is less than or equal to 5.

(c) Represent the occurrence sequences from (b) in a directed graph which has a node for each marking and an arc for each step. Such a graph is called an occurrence graph (to which we shall return in Chap. 5).

Exercise 2.5.
Consider the resource allocation system from Sect. 1.2.

(a) Remove the cycle counter from the CP-net in Fig. 1.7. This is done by replacing P with U, (x,i) with x, (p,i+1) with p, (q,i+1) with q, (p,0) with p and (q,0) with q.

(b) Show that the CP-net from (a) and the PT-net in Fig. 1.3 are behaviourally equivalent, in the sense that there is a one-to-one correspondence between their sets of occurrence sequences. This can be done by showing that the translation in Def. 2.15 maps the CP-net from (a) into a PT-net that is nearly identical to the PT-net in Fig. 1.3.

(c) Find a pair of partitions which make it possible to map the PT-net in Fig. 1.3 into a CP-net with the same net structure as Fig. 1.7. Discuss the difference between the two CP-nets.

(d) Find a pair of partitions which make it possible to map the PT-net in Fig. 1.3 into a CP-net with the same net structure as Fig. 1.11.

(e) Find a pair of partitions which make it possible to map the PT-net in Fig. 1.3 into a CP-net with the same net structure as Fig. 1.12.

Exercise 2.6.
Consider the distributed data base system from Sect. 1.3.

(a) Translate the CP-net in Fig. 1.13 into an equivalent PT-net.

Exercise 2.7.
Consider the model railway from Exercise 1.5.

(a) Translate the two CPN diagrams of Exercise 1.5 (b)–(c) into equivalent PT-nets. Compare the two nets with the PT-net of Exercise 1.5 (a).

Chapter 3

Hierarchical Coloured Petri Nets

In Chaps. 1 and 2 we presented non-hierarchical CP-nets. In this chapter we shall see how these can be extended to hierarchical nets, i.e., how it is possible to construct a large CP-net by combining a number of smaller nets. A hierarchical CP-net can be constructed top-down, bottom-up, or by mixing these two strategies.

Looking at the history of high-level programming languages, it is obvious that their success depends to a very large degree upon the existence of subroutines and modules – making it possible to construct a large description from smaller units which can be investigated more or less independently of each other. The absence of compositionality has been one of the main critiques raised against Petri net models. To meet this critique, hierarchical CP-nets have been developed that make it possible to relate a number of individual CP-nets to each other in a formal way, i.e., in a way which has a well-defined semantics and thus allows formal analysis.

This chapter introduces two of the language constructs by which hierarchical CP-nets can be constructed. The two hierarchy constructs are known as substitution transitions and fusion places. They allow the user to build a large hierarchical CP-net by composing a number of smaller non-hierarchical CP-nets. Moreover, we investigate the relationship between hierarchical CP-nets and non-hierarchical CP-nets, and it turns out that each hierarchical CP-net can be translated into a behaviourally equivalent non-hierarchical CP-net, and vice versa. The importance of this translation is similar to that of the translation between non-hierarchical CP-nets and PT-nets. It tells us how to generalize the basic concepts and the analysis methods of non-hierarchical nets to hierarchical nets.

Section 3.1 introduces substitution transitions and fusion places. This is done by means of a small example, which describes a simple ring protocol. Section 3.2 contains a second example of hierarchical nets, describing a telephone system. Section 3.3 defines the structure of hierarchical CP-nets, while Sect. 3.4 defines the behaviour. Section 3.5 investigates the relationship between hierarchical CP-nets and non-hierarchical CP-nets; the section is theoretical and can be skipped by readers who are interested primarily in the practical application of CP-nets.

3.1 Introduction to Hierarchical CP-nets

It is important to understand that the basic idea behind hierarchical CP-nets is to allow the modeller to construct a large model by combining a number of small CP-nets into a larger net. This is similar to the situation in which a programmer constructs a large program from a set of modules and subroutines. However, the idea is different from those approaches which relate two or more separate subnets to each other – in order to compare their behaviour – but *without* combining them into a single net. Such approaches are analogous to program transformations, and the individual subnets are alternative descriptions of the same system.

As mentioned in the introduction to this chapter, it is always possible to translate a hierarchical CP-net into a non-hierarchical CP-net – which in turn can be translated into a PT-net. This means that the theoretical modelling powers of these three classes of nets are the same. However, from a practical point of view, the three net classes have very different properties. To cope with large systems we need to develop strong structuring and abstraction concepts. The first very substantial step on this path was to replace low-level Petri nets with high-level nets. The second step is to introduce hierarchical nets. In terms of programming languages, the first step can be compared to the introduction of types – allowing the programmer to work with structured data elements instead of single bits. The second step may then be compared to the development of programming languages with subroutines and modules – allowing the programmer to construct a large model as a set of smaller models which are related to each other in a well-defined way. From a theoretical point of view, machine languages (or even Turing machines) are equivalent to the most powerful modern programming languages. From a practical point of view, this is of course not the case. One of the most important limitations that system developers face today is their own inability to cope with many details at the same time. In order to develop and analyse complex systems they need structuring and abstraction concepts that allow them to work with a selected part of the model without being distracted by the low-level details of the remaining parts. Hierarchical nets provide the Petri net modeller with such abstraction mechanisms.

The intention has been to make a set of hierarchy constructs which is general enough to be used with many different development methods and with many different analysis techniques. This means that we present the hierarchy constructs *without* prescribing specific methods for their use. Such methods have to be developed and written down – but this can only be done as we get more experience with the practical use of the hierarchy constructs. Eventually the new development methods and analysis techniques will influence the definition of the hierarchy constructs in the same way that modern programming languages have been influenced by the progress in the areas of programming methodology and verification techniques. During the design of the hierarchy constructs we have, of course, been influenced by the constructs and methods used with other graphical description languages and with modern programming languages.

The concept of hierarchical nets is much younger than the concept of high-level nets, and this means that the hierarchy concepts are likely to undergo many improvements and refinements (in the same way that the first very simple concept of subroutines has undergone dramatic changes to become the procedure concept of modern programming languages). In other words, we do not claim that our current proposal will be the "final solution". However, we do think that it constitutes a good starting point for further research and practical experiences in the area of hierarchical nets. In Chap. 7 we describe a number of industrial applications of hierarchical CP-nets.

Substitution of transitions

The intuitive idea behind substitution transitions is to allow the user to relate a transition (and its surrounding arcs) to a more complex CP-net – which usually gives a more precise and detailed description of the activity represented by the substitution transition. The idea is analogous to the hierarchy constructs found in many graphical description languages (e.g., data flow diagrams) and it is also, in some respects, analogous to the module concepts found in many modern programming languages. At one level, we want to give a simple description of the modelled activity without having to consider internal details about how it is carried out. At another level, we want to specify the more detailed behaviour. Moreover, we want to be able to integrate the detailed specification with the more crude description – and this integration must be done in such a way that it becomes meaningful to speak about the behaviour of the *combined* system. Now let us consider a small example, consisting of a simple ring network with four different sites. The purpose of the example is to explain the semantics of substitution transitions, and thus the described network is far too simple to be realistic.

As mentioned above, we want to relate individual CP-nets to nodes, which are members of other CP-nets, and this means that our description will contain a *set* of non-hierarchical CP-nets – which we shall call **pages**. Figure 3.1 shows a page from the network system. The page has a **page name** *NetWork* (which is a text string) and a **page number** *10* (which is a non-negative integer). The page name and the page number do not have to be unique and they may be missing. Their only purpose is to make it possible for human beings (and computer systems) to refer to the page. This is usually done by specifying both the name and the number: *NetWork#10* – but it is also possible to leave one of them out and only write: *NetWork* or *10*.

The page *NetWork#10*, which is shown in Fig. 3.1, has four places and four transitions – positioned in a ring. The four transitions are **substitution transitions**. This can be seen because each of them has a small HS-tag adjacent to it (HS ≈ Hierarchy + Substitution). The text strings in the dashed boxes next to the HS-tags are called **hierarchy inscriptions** and they define the details of the substitutions. The notation with the HS-tags and the hierarchy inscriptions are adopted from the CPN tools described in Chap. 6, and they work in a way similar to the small circles and text strings by which we represent the current markings of a CP-net. This means that a double-click on the HS-tag makes the hierarchy inscription (which may be quite large) visible/invisible. In this way it is possible to have a CP-net with a complex hierarchical structure without getting the individual pages too overloaded.

The first line of each hierarchy inscription tells us the identity of the **subpage**, i.e., the page which contains the detailed description of the activity modelled by the corresponding substitution transition. Each substitution transition is said to be a **supernode** (of the corresponding subpage) while the page of a substitution transition is a **superpage** (of the corresponding subpage). In our example, we can see that all four substitution transitions of *NetWork#10* share the same subpage *Site#11* which is shown in Fig. 3.2. This means that the hierarchical net will have four instances of *Site#11*. Each of these **page instances** will have its own private marking, which is independent of the markings of the other instances (in a similar way that each procedure call has its own private copies of the local variables in the procedure).

When a CP-net is simulated by means of the CPN simulator described in Chap. 6, we have a window for each page. The window shows the marking of one page instance at a time, and it is possible for the user to switch from one instance to another. We will return to the remaining lines of the hierarchy inscriptions in a moment, but let us first take a look at *Site#11* which describes an individual site in the ring network.

The declarations in the middle of Fig. 3.2 declares a constant NoOfSites = 4, which tells us how many sites the ring network has. Moreover, three colour sets are declared: INT contains all integers. SITES contains the elements {S(1),S(2), …,S(NoOfSites)}, which are used to identify the individual sites of the network. Finally, PACK describes the individual packages which are sent in the network.

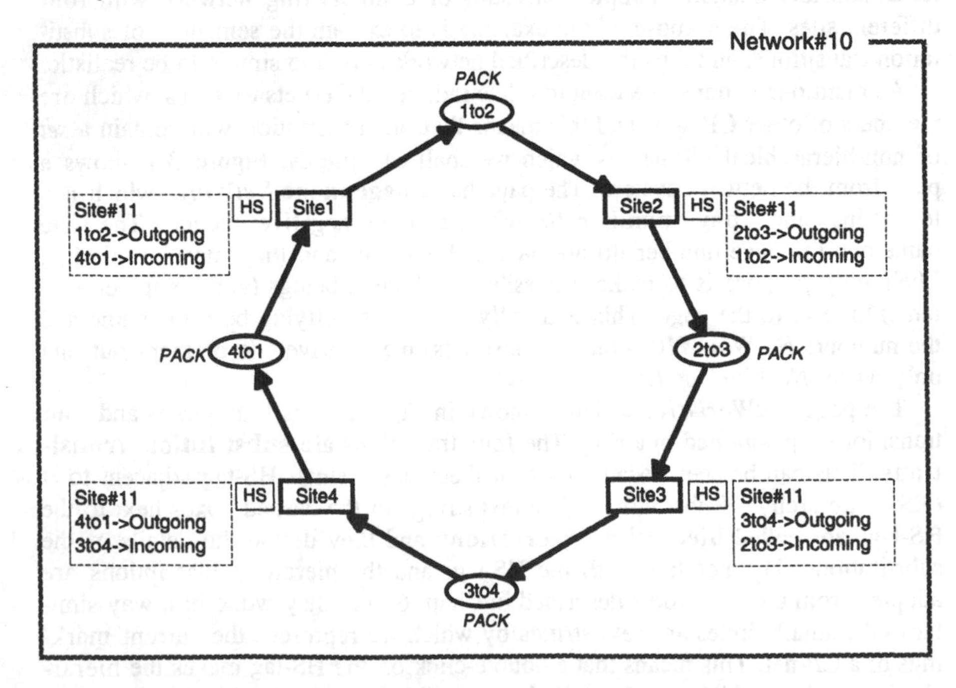

Fig. 3.1. *NetWork#10* describes a ring network with four different sites

The format is a record, containing an se-field for the identity of the sender, an re-field for the identity of the receiver and a no-field for a package number. The actual data content of the packages is ignored, but if desired it could of course easily be added as an additional field of the record.

Site#11 has three different transitions. Each occurrence of *NewPack* creates a new package. The no-field of the new package is determined by the token on the place *PackNo* (and the colour of this token is increased by one so that the next package will get a package number which is one higher). The se-field of the new package is determined by the expression S(inst()) – which uses the predeclared function inst() to refer to the page instance number of the page instance to which the occurring transition belongs. Allowing guards, arc expressions and initialization expressions to be dependent on the page instance number has turned out to be extremely useful – it often makes a hierarchical CP-net much more readable. By convention we use consecutive page instance numbers, starting from one. In the present example we assume that the page instance of transition *SiteX* has X as page instance number. Finally, the re-field is determined by the vari-

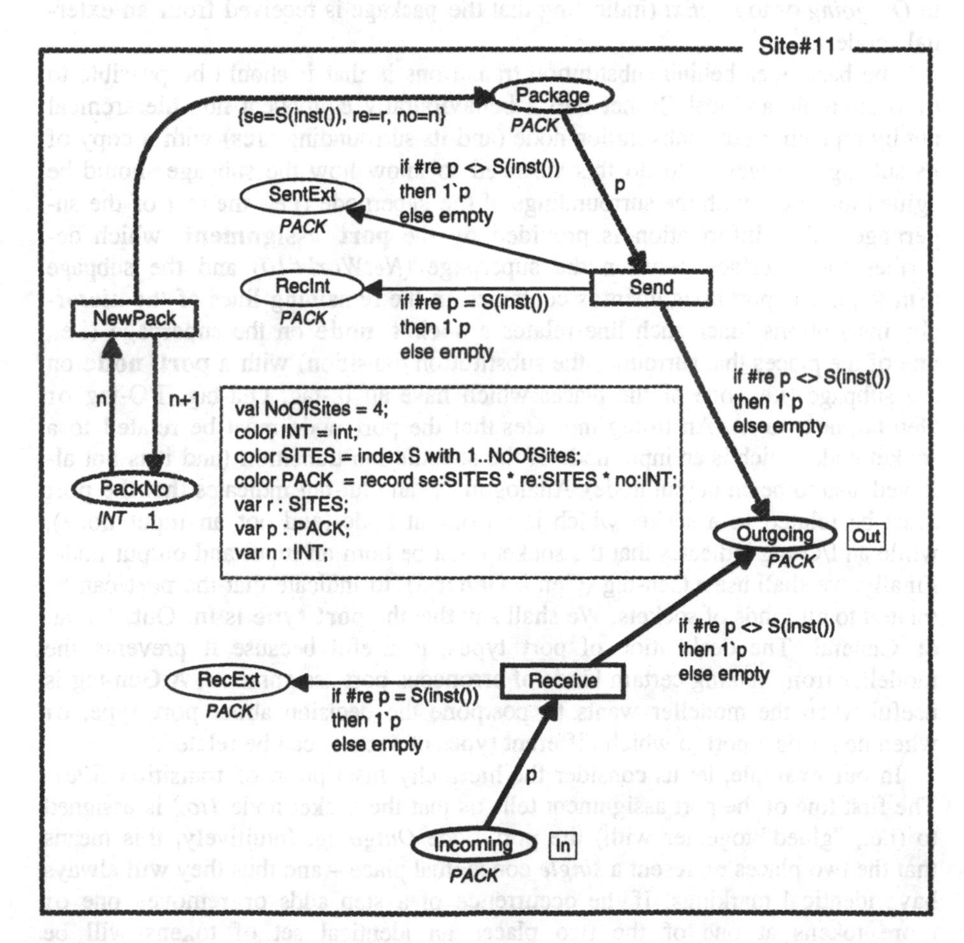

Fig. 3.2. ***Site#11*** **describes an individual site of the ring network**

able r – which does not appear in the guard and input arc expressions. This means that the binding of r does not influence the enabling, and thus r can take an arbitrary value (from SITES). This means that a site can send a package to itself, but if desired this could of course easily be prevented by adding a guard specifying that r ≠ S(inst()).

The created packages are handled by the transition *Send*, which inspects the re-field of the package. This is done by means of the expression *#re p* which denotes the re-field of the record p. When the re-field indicates that the receiver is different from the present site, the package is transferred to the place *Outgoing* (which is the "output gate" to the rest of the network) and a copy of the package is put on the place *SentExt* (indicating that the package is sent to an external receiver). Otherwise the package is sent directly to the place *RecInt* (indicating that the package is received from an internal sender).

Finally, the transition *Receive* inspects all the packages which arrive at the place *Incoming* (which is the "input gate" from the rest of the network). Again the re-field is inspected, and based on this inspection the package is routed, either to *Outgoing* or to *RecExt* (indicating that the package is received from an external sender).

The basic idea behind substitution transitions is that it should be possible to translate a hierarchical CP-net into a behaviourally equivalent non-hierarchical net by replacing each substitution node (and its surrounding arcs) with a copy of its subpage. However, to do this we need to know how the subpage should be "glued together" with the surroundings of the supernode (i.e., the rest of the superpage). This information is provided by the **port assignment**, which describes the interface between the superpage (*NetWork#10*) and the subpage (*Site#11*). The port assignment is contained in the remaining lines of the hierarchy inscriptions. Each such line relates a **socket node** on the superpage (i.e., one of the places that surrounds the substitution transition) with a **port node** on the subpage (i.e., one of the places which have an In-tag, Out-tag, I/O-tag or Gen-tag next to it). An In-tag indicates that the port node must be related to a socket node which is an input node of the substitution transition (and it is not allowed also to be an output node). Analogously, an Out-tag indicates that the port must be related to a socket which is an output node (and not an input node), while an I/O-tag indicates that the socket must be both an input and output node. Finally, we shall use a Gen-tag (Gen ≈ General) to indicate that the port can be related to all kinds of sockets. We shall say that the **port type** is In, Out, In/Out or General. The declaration of port types, is useful because it prevents the modeller from making certain kinds of erroneous port assignments. A Gen-tag is useful when the modeller wants to postpone the decision about port type, or when he needs a port to which different types of sockets can be related.

In our example, let us consider the hierarchy inscription of transition *Site1*. The first line of the port assignment tells us that the socket node *1to2* is assigned to (i.e., "glued" together with) the port node *Outgoing*. Intuitively, this means that the two places represent a *single* conceptual place – and thus they will always have identical markings. If the occurrence of a step adds or removes one or more tokens at one of the two places an identical set of tokens will be added/removed at the other. Analogously, the second line tells us that the other socket node *4to1* is assigned to the other port node *Incoming*. The remaining three hierarchy inscriptions (of *Site2*, *Site3* and *Site4*) are interpreted in a similar way, and intuitively this tells us that the hierarchical CP-net with the two pages *NetWork#10* and *Site#11* is behaviourally equivalent to the non-hierarchical CP-net in Fig. 3.3. For clarity we have omitted all the net inscriptions in Fig. 3.3. They are identical to the corresponding net inscriptions of Figs. 3.1 and 3.2 (except that inst() is replaced by 1, 2, 3 and 4, respectively). Notice that the place *1to2* has been formed by gluing together three different instances of places: *1to2* from *NetWork#10*, *Outgoing* from the *first* instance of *Site#11* and *Incoming* from the *second* instance of *Site#11*. Similar remarks apply to the other three places in the ring, i.e., *2to3*, *3to4* and *4to1*.

The translation from hierarchical CP-nets to non-hierarchical CP-nets – obtained by replacing substitution nodes with their subpages – plays a role similar to the translation from non-hierarchical CP-nets to PT-nets. This means that the only purpose of the translation is to learn how to define the semantics and the behavioural properties of hierarchical nets. It is important to understand that we never make the translation for a particular hierarchical CP-net. When we describe a large system we use hierarchical CP-nets directly, without constructing

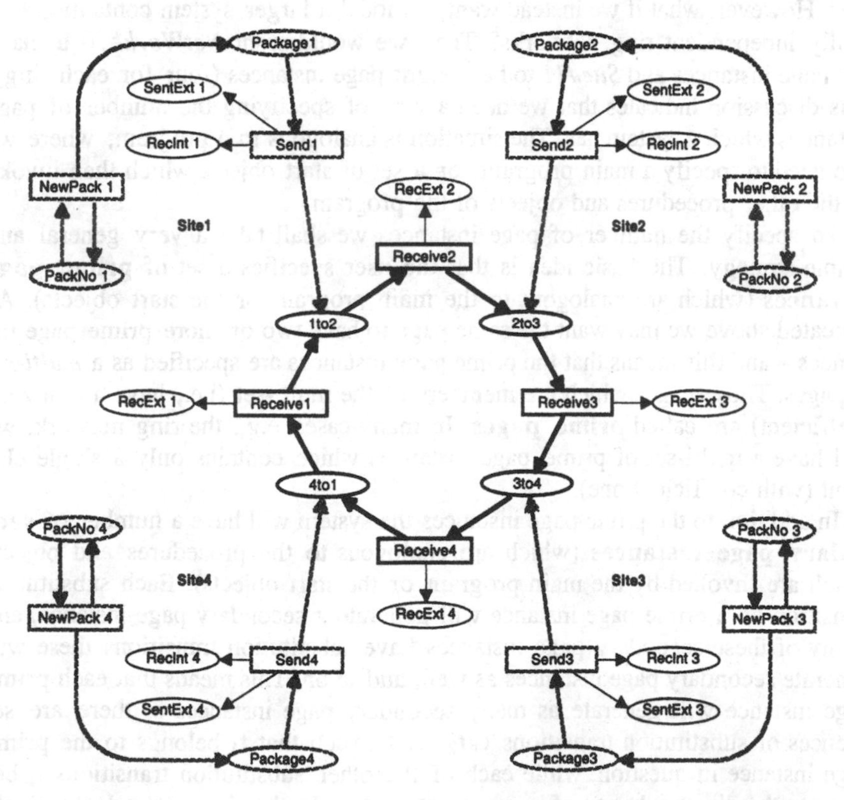

Fig. 3.3. Non-hierarchical CP-net with the same behaviour as the hierarchical CP-net that contains the pages *NetWork#10* and *Site#11*

the non-hierarchical equivalent. Analogously, we analyse a hierarchical CP-net directly, without constructing the non-hierarchical equivalent.

In the ring network we have only one level of substitution. However, in general it is necessary to allow subpages to have substitution transitions, and thus subpages. We then have to be a little more precise with our terminology, and we shall say that a page s* with a substitution transition t related to the page s is a **direct superpage** (of s), while t is a **direct supernode** (of s), and s a **direct subpage** (of s* and t). Next, we extend these concepts by means of transitivity, and we say that s* is a **superpage** (of s), t a **supernode** (of s), and s a **subpage** of (s* and t), iff there exists a non-empty sequence of substitution transitions t_1 t_2 ... t_m such that $t=t_1$ is a substitution transition on s*, while each of the other substitution transitions t_i belongs to the direct subpage of its predecessor t_{i-1}, and s is the direct subpage of t_m.

Page instances

In the ring network we have assumed that there is only a single page instance of *NetWork#10*, and that the only page instances of *Site#11* are the four which exist because of the four substitution transitions on *NetWork#10*. This is a rather obvious choice – as long as we want to model a single ring with four individual sites. However, what if we instead want to model a larger system containing two totally independent ring networks? Then we would want *NetWork#10* to have two page instances and *Site#11* to have eight page instances (four for each ring). This discussion indicates that we need a way of specifying the number of page instances which a system has. The situation is analogous to a program, where we also need to specify a main program, or a set of start objects which then invoke all the other procedures and objects of the program.

To specify the number of page instances we shall take a very general and simple strategy. The basic idea is that the user specifies a set of **prime page instances** (which are analogous to the main program or the start objects). As indicated above we may want the same page to have two or more prime page instances – and this means that the prime page instances are specified as a *multi-set* of pages. Those pages which are members of the multi-set (i.e., have a non-zero coefficient) are called **prime pages**. In many cases, e.g., the ring network, we will have a multi-set of prime page instances which contains only a single element (with coefficient one).

In addition to the prime page instances the system will have a number of **secondary page instances** (which are analogous to the procedures and objects which are invoked by the main program or the start objects). Each substitution transition on a prime page instance will generate a secondary page instance, and if any of these secondary page instances have substitution transitions these will generate secondary page instances as well, and so on. This means that each prime page instance will generate as many secondary page instances as there are sequences of substitution transitions t_1 t_2 ... t_m such that t_1 belongs to the prime page instance in question, while each of the other substitution transitions t_i belongs to the direct subpage of its predecessor t_{i-1}. In the ring network the single prime page instance generates four secondary page instances – corresponding to the following four sequences (each of which consists of only a single substitution transition): Site1; Site2; Site3; Site4.

To give an overview of the set of pages, the multi-set of prime and secondary page instances, and the superpage/subpage relationships in a hierarchical CP-net, we use a **page hierarchy graph**. This is a directed graph which contains a node for each page and an arc for each direct superpage/subpage relationship. Each node is inscribed by the name and number of the corresponding page, while each arc is inscribed with the names of the corresponding substitution transitions (i.e., the names of those substitution transitions which belong to the page of the source node, and have the page of the destination node as direct subpage). The set of prime page instances is indicated by positioning the word "Prime" next to each prime page, adding a number, e.g., "Prime : 3" if the coefficient differs from one. The page hierarchy graph for the ring network is shown in Fig. 3.4. Notice that the arc has four inscriptions (because NetWork#10 has four substitution transitions).

The page hierarchy graph can also be used to find the number of page instances of a given page. This can be calculated as the number of different "routes" which leads from an appearance of a prime page instance to the page in question (when an arc represents more than one substitution transition there are routes for each of these). Routes of length zero represent prime page instances, while all other routes represent secondary page instances.

Fusion of places

The intuitive idea behind fusion of places is to allow the user to specify that a set of places are considered to be identical, i.e., they all represent a single conceptual place even though they are drawn as individual places. This means that when a token is added/removed at one of the places, an identical token will be added/removed at all the others. The relationship between the members of a fusion set is, in some respects, similar to the relationship between two places which are assigned to each other by a port assignment – this will become much clearer in Sect. 3.3 (when we define place instance groups).

The places that participate in such a **fusion set** may belong to a single page or to several different pages. When all members of a fusion set belong to a single page and that page only has one page instance, place fusion is nothing other than a drawing convenience that allows the user to avoid too many crossing arcs. However, things become much more interesting when the members of a fusion set belong to several different pages *or* to a page that has several page instances. In that case fusion sets allow the user to specify a behaviour which it may be cumbersome to describe without fusion.

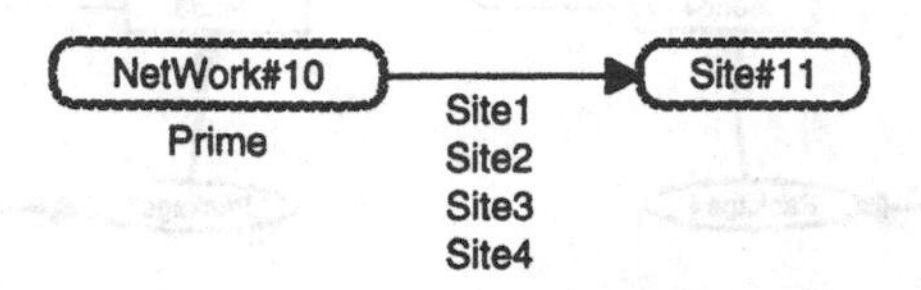

Fig. 3.4. Page hierarchy graph for the ring network

There are three different kinds of fusion sets: **global fusion sets** are allowed to have members from many different pages, while **page fusion sets** and **instance fusion sets** only have members from a single page. The difference between the last two is the following. A page fusion unifies all the instances of its places (independently of the page instance at which the place instance appear), and this means that the fusion set only has one "resulting place" which is "shared" by all instances of the corresponding page. In contrast, an instance fusion set only identifies place instances that belong to the *same* page instance, and this means that the fusion set has a "resulting place" for each page instance. The semantics of a global fusion set is analogous to that of a page fusion set – in the sense that there only is one "resulting place" (which is common for all instances of all the participating pages). To allow modular analysis of hierarchical CP-nets, global fusion sets should be used with care.

The difference between page and instance fusion sets can easily be illustrated by the ring network. In Fig. 3.5 we show how the non-hierarchical equivalent in Fig. 3.3 is modified when we define two fusion sets on *Site#11:* an instance fusion set containing the two places *RecExt* and *RecInt*, and a page fusion set containing a single place *SentExt*. The first fusion set combines *RecExt* and *RecInt* into a single place – and since we deal with an instance fusion set, we get a "resulting place" for each of the four page instances of *Site#11*. The second fu-

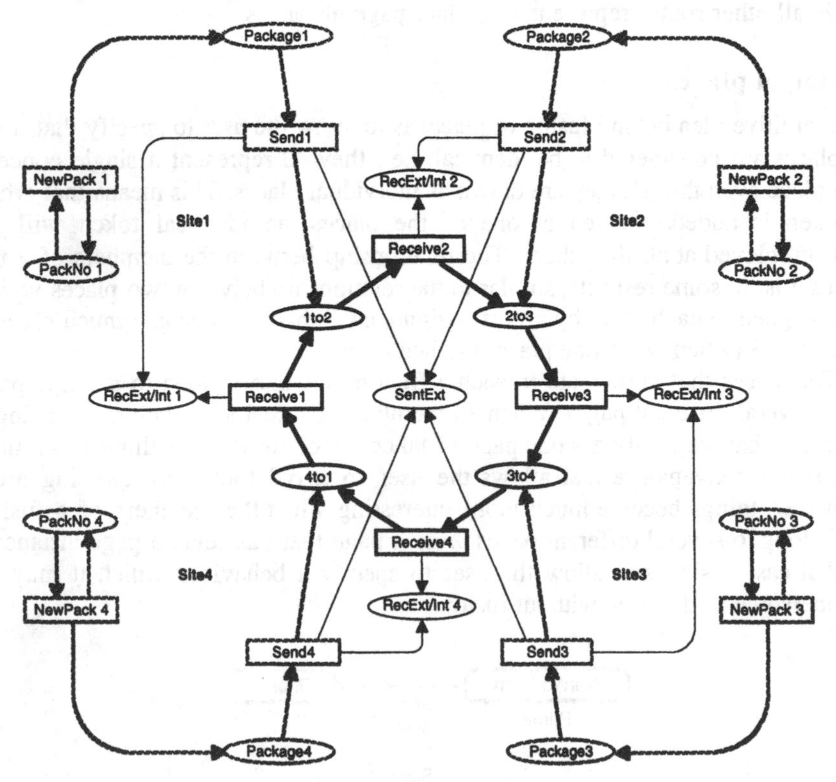

Fig. 3.5. Non-hierarchical equivalent of a hierarchical CP-net with two fusion sets

sion set combines *SentExt* with itself – but now we deal with a page fusion set, and thus there is only one "resulting place" shared by all four page instances.

Above we have illustrated the difference between page and instance fusion sets by drawing a non-hierarchical CP-net which is behaviourally equivalent to our hierarchical CP-net. However, it should again be mentioned that hierarchical CP-nets form a modelling language in their own right, and in practice we never construct the non-hierarchical equivalent of a particular hierarchical CP-net.

3.2 Example of Hierarchical CP-nets: Telephones

This section contains a second example of hierarchical CP-nets. We describe the public telephone system – as it is conceived by a user (and not by a telephone technician). We ignore time-outs and special services such as conference calls etc. The system was considered in Sect. 1.5 and in Exercise 1.8.

The telephone system has a single prime page *Phone#1* (with coefficient 1). This page is shown in Fig. 3.6, and it has three transitions which all are substitution transitions. The substitution transitions are drawn with additional line thickness. However, we have omitted the hierarchy inscriptions because, in this particular system, they would only show that each substitution transition has a subpage with the same name, and that each socket node is assigned to a port node with the same name. From *Phone#1* we can see that the activities of the phone system are divided into three parts: the *Establishment of a Connection*, the *Breaking of a connection by the Sender*, and the *Breaking of a connection by the Recipient*.

The telephones are represented by U-tokens which at *Phone#1* only can be in two different states, *Inactive* and *Connected*. The telephone exchange (i.e., the electronics of the telephone switch) is represented by a single place, *Connection*. This place has a number of U×U–tokens, and each of them represents an established connection. The first element in a U×U–token identifies the **sender** (i.e., the phone from which the call was made), while the second element identifies the **recipient** (i.e., the phone which was called). It is important not to confuse the recipient with the receiver, which is the part of the telephone which you put to your ear. The receiver can be lifted and replaced, and we shall be a bit sloppy and say that the receiver is lifted or replaced by the sender/recipient – where we actually mean that a *person* at the sender/recipient performs the lifting/replacement operation. Analogously, we shall, e.g., say that the sender dials a number.

Next, let us take a closer look at the *Establishment of a Connection*, i.e., the page *EstabCon#2* in Fig. 3.6. This page has two substitution transitions, describing how the *Dialing of a number* and the *Breaking of a connection due to Engaged recipient* are performed. The telephones can now be in four additional states:

- *Engaged*, which intuitively is the opposite of *Inactive*. Notice that each transition which adds (removes) a token at *Inactive*, removes (adds) an identical token at *Engaged*. This property is not true for the substitution transitions, but this does not matter, because the arc expressions of the substitution transitions do not influence the behaviour of the hierarchical CP-net. We shall return to the arc expressions of the substitution transitions later in this section.

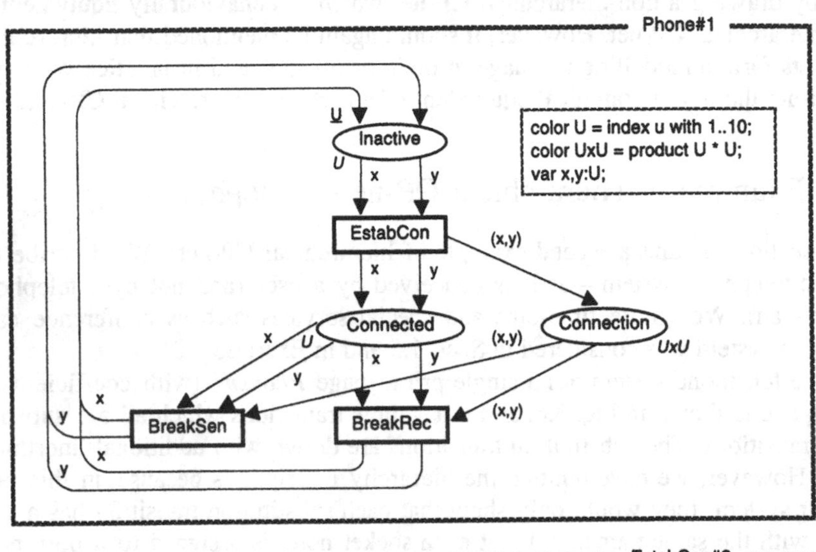

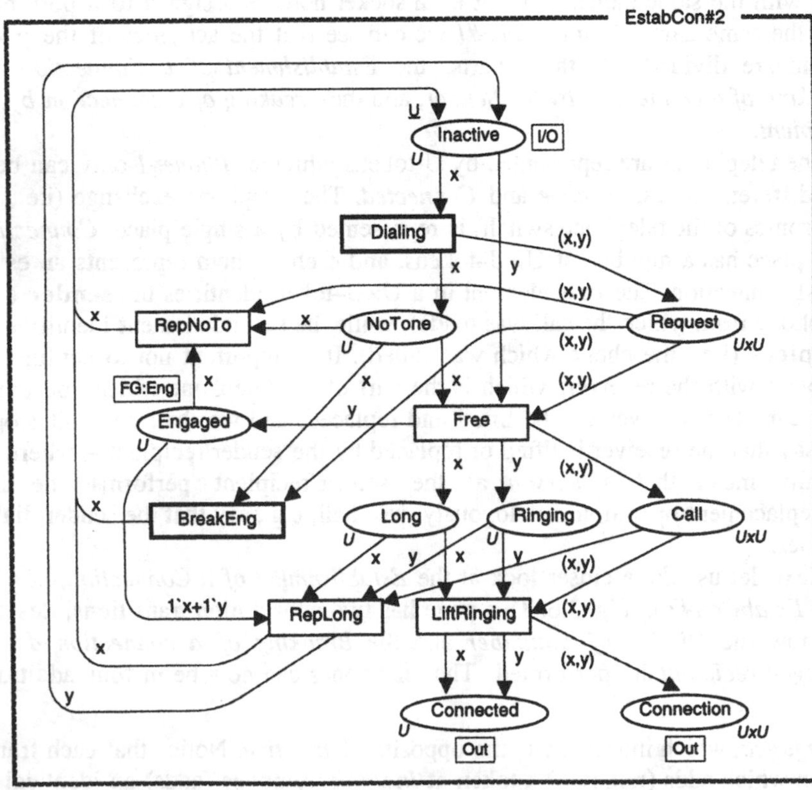

Fig. 3.6. Two pages of the hierarchical telephone system

The property above, and the initialization expressions of *Inactive* and *Engaged*, tell us that each telephone always has a token either at *Inactive* or at *Engaged*. Notice that *Engaged* is a member of a global fusion set (FG ≈ Fusion + Global). This fusion set is called *Eng*, and it has five different places as members, which all are called *Engaged* and belong to five different pages.

- *NoTone*, which represents the state where no tone is heard at the sender in the few seconds where the system checks whether the recipient is *Inactive* or *Engaged*.

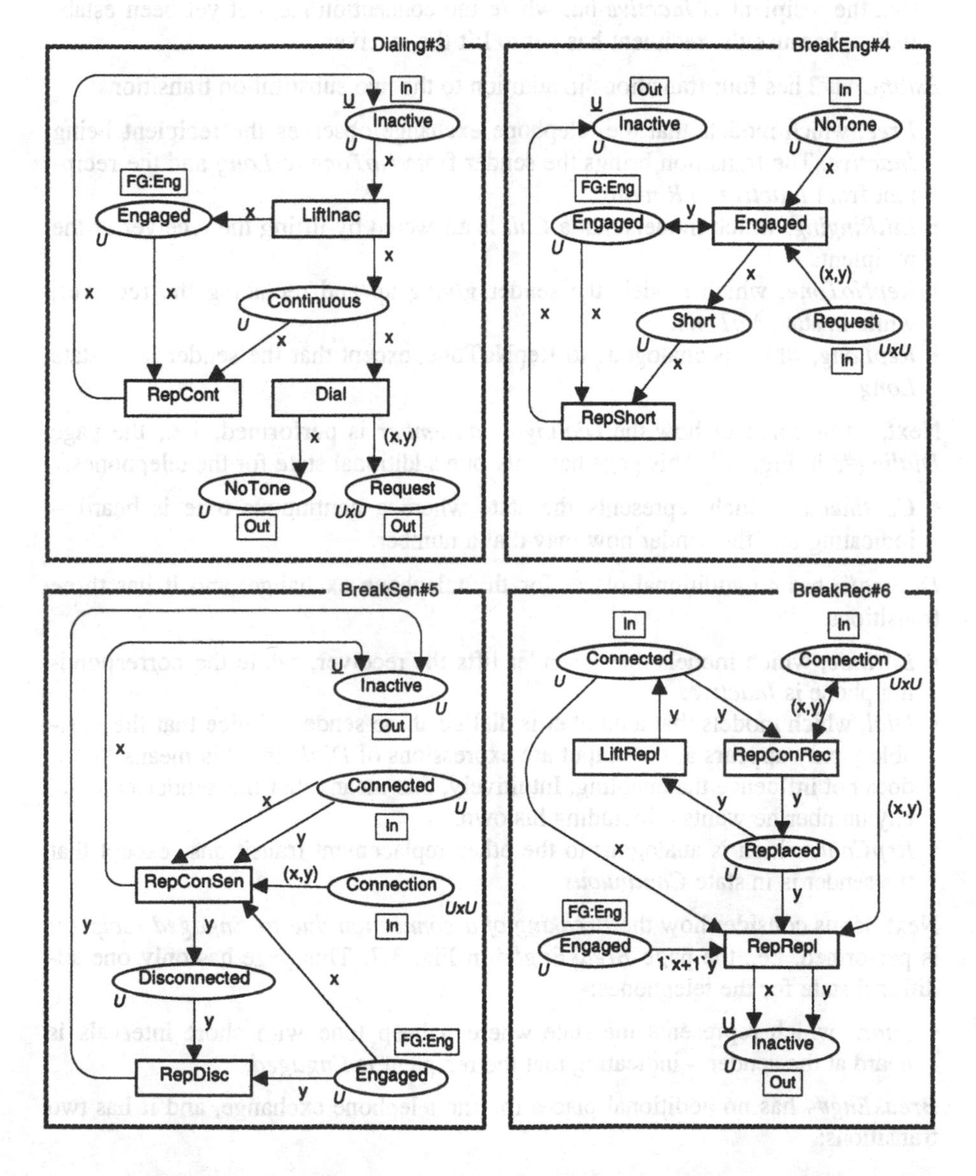

Fig. 3.7. The additional four pages of the hierarchical telephone system

- *Long*, which represents the state where a beep tone with long intervals is heard at the sender – indicating that the recipient is *Inactive*.
- *Ringing*, which represents the state where the recipient is ringing, while the sender is in state *Long*.

EstabCon#2 has two additional places for the telephone exchange:

- *Request*, which represents demands for connections which have just been made, but not yet been established.
- *Call*, which represents demands for connections – where it has been checked that the recipient is *Inactive* but where the connection has not yet been established, because the recipient has yet to lift the receiver.

EstabCon#2 has four transitions in addition to the two substitution transitions:

- *Free*, which models that the telephone exchange observes the recipient being *Inactive*. The transition brings the sender from *NoTone* to *Long* and the recipient from *Inactive* to *Ringing*.
- *LiftRinging*, which models that a *Call* is answered by lifting the receiver at the recipient.
- *RepNoTone*, which models the sender giving up and replacing the receiver, while in state *NoTone*.
- *RepLong*, which is analogous to RepNoTone, except that the sender is in state *Long*.

Next, let us consider how the *Dialing of a number* is performed, i.e., the page *Dialing#3* in Fig. 3.7. This page has only one additional state for the telephones:

- *Continuous*, which represents the state where a continuous tone is heard – indicating that the sender now may dial a number.

Dialing#3 has no additional places for the telephone exchange, and it has three transitions:

- *LiftInac*, which models that a sender lifts the receiver, while the corresponding phone is *Inactive*.
- *Dial*, which models that a number is dialled at the sender. Notice that the variable y only appears at the output arc expressions of *Dial*, and this means that it does not influence the enabling. Intuitively, this means that the sender can dial any number he wants – including his own.
- *RepCont*, which is analogous to the other replacement transitions, except that the sender is in state *Continuous*.

Next, let us consider how the *Breaking of a connection due to Engaged recipient* is performed, i.e., the page *BreakEng#4* in Fig. 3.7. This page has only one additional state for the telephones:

- *Short*, which represents the state where a beep tone with short intervals is heard at the sender – indicating that the recipient is *Engaged*.

BreakEng#4 has no additional places for the telephone exchange, and it has two transitions:

- *Engaged*, which models that the telephone exchange observes the recipient being *Engaged*. The transition brings the sender from *NoTone* to *Short* while the state of the recipient is unaltered.
- *RepShort*, which is analogous to the other replacement transitions, except that the sender is in state *Short*.

Next, let us consider how the *Breaking of a connection by the Sender* is performed, i.e., the page *BreakSen#5* in Fig. 3.7. This page has only one additional state for the telephones:

- *Disconnected*, which represents the state where the recipient has become disconnected because the sender has broken the connection by replacing the receiver.

BreakSen#5 has no additional places for the telephone exchange, and it has two transitions:

- *RepConSen*, which models that the sender breaks the connection by replacing the receiver. Notice that the effect of this transition is *not* totally analogous to the other replacement transitions, because it is only the sender that returns to *Inactive*, while the recipient becomes *Disconnected*.
- *RepDisc*, which is analogous to the other replacement transitions, except that the recipient is in state *Disconnected*.

Finally, let us consider how the *Breaking of a connection by the Recipient* is performed, i.e., the page *BreakRec#6* in Fig. 3.7. This page has only one additional state for the telephones:

- *Replaced*, which represents the state where the recipient has replaced the receiver. Notice that this action – at least in the Danish telephone system – does not terminate the connection, and this means that the conversation can be continued, if the recipient again lifts the receiver (for more information, see the description of *RepConRec* and *LiftRepl* below).

BreakRec#6 has no additional places for the telephone exchange, and it has three transitions:

- *RepConRec*, which models that the recipient replaces the receiver. Notice that the effect of this transition is *not* analogous to the other replacement transitions, because neither the sender nor the recipient returns to *Inactive*. Instead the sender remains *Connected*, while the recipient becomes *Replaced*.
- *LiftRepl*, which represents the fact that the recipient lifts the receiver, while the recipient is in state *Replaced*.
- *RepRepl*, which is analogous to the other replacement transitions, except that the sender is in state *Connected* while the recipient is in state *Replaced*.

The telephone system has the page hierarchy graph shown in Fig. 3.8 which shows that each page has exactly one page instance. The telephone system has the non-hierarchical equivalent shown in Fig. 3.9, where we have omitted all the arcs updating the marking of *Engaged*. This place has (in addition to the double-headed arc shown) exactly the same arcs and arc expressions as *Inactive*, but with opposite directions. Moreover, we have used different kinds of highlighting to show the states/actions of the sender (thick black lines), the recipient (thick shaded lines) and the telephone exchange (thin lines). The same idea could have been used for the hierarchical version.

Now let us take a closer look at the arc expressions which surround the substitution transitions. These arc expressions do not influence the behaviour of the hierarchical CP-net, because the substitution transitions and their surrounding arcs are replaced by the subpages, and this means that it does not make sense to talk about an enabled or an occurring substitution transition. Thus we can omit the arc expressions surrounding the substitution transitions, as we did for the ring network. This is considered to be a shorthand for empty (cf. the remarks for Def. 2.5 (viii)). However, it is also possible to specify these arc expressions in a less trivial way – and they then act rather like comments, giving the reader of the hierarchical CP-net an idea about the behaviour of the subpages. The CPN simulator described in Chap. 6 allows the modeller to specify that the subpage of a given substitution transition should be temporarily ignored. Then the transition behaves as an ordinary transition – and the guard and arc expressions become significant. By using this facility, and by changing the multi-set of prime pages, it is easy to debug selected parts of a large hierarchical CP-net without having to "cut them out" of the CP-net.

In the telephone example the arc expressions of the substitution transitions are used in two slightly different ways. The arc expressions of *BreakSen@Phone#1* give a rather precise description of the behaviour of the subpage *BreakSen#5*. The only differences are that the arc expressions (of course) do not show that the activity is executed in two subsequent steps, and that the arc expressions do not show the updating of *Engaged* (which does not exist at the abstraction level of *Phone#1)*. The arc expressions of *EstabCon@Phone#1* are used in a different way. These arc expressions only describe the standard case, i.e., the case in which a *Connection* is established, and they ignore the possibilities where the recipient is *Engaged* or the sender replaces the receiver. It would not have been particularly difficult to include these additional possibilities into the arc expressions. However, the modeller has chosen, at the abstraction level of Phone#1, to concentrate on the standard case.

It may be argued that it is necessary to force the modeller to have a one-to-one correspondence between the arc expressions of a substitution transition and the behaviour of the corresponding subpage – because this will make it easier to develop methods allowing modular analysis. We think it is very important to develop such incremental analysis methods, built upon strict behavioural equivalence between the different levels of description. However, as indicated above,

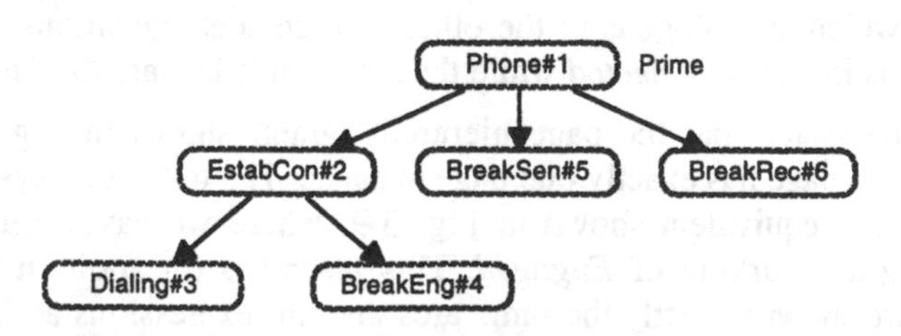

Fig. 3.8. Page hierarchy graph for the telephone system

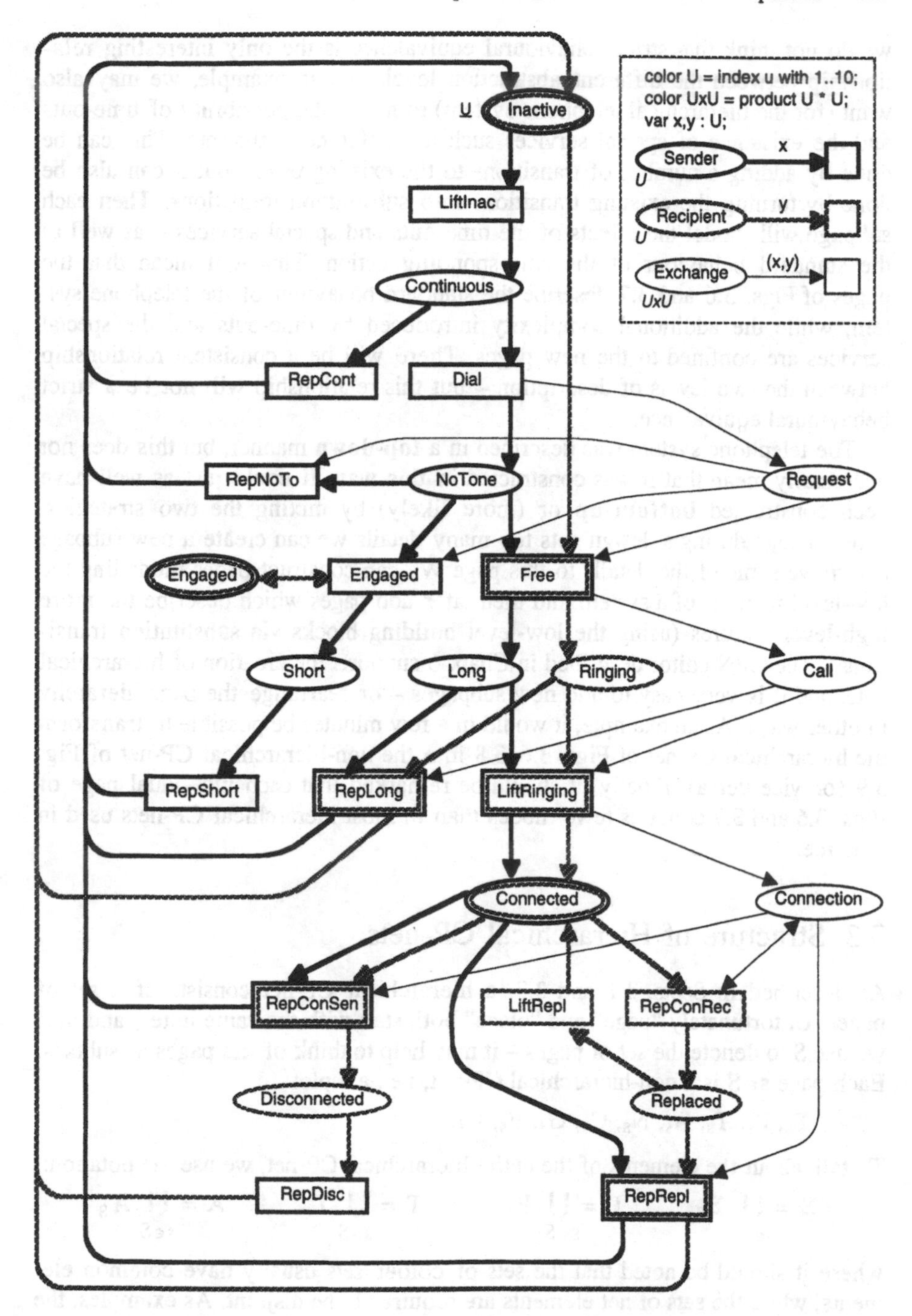

Fig. 3.9. Non-hierarchical equivalent for the telephone system

we do not think that strict behavioural equivalence is the only interesting relationship between the different abstraction levels. As an example, we may also want (for the hierarchical telephone system) to model the possibility of time-outs and the existence of special services such as conference calls, etc. This can be done by adding a number of transitions to the existing pages, but it can also be done by turning the existing transitions into substitution transitions. Then each subpage will model the effects of the time-outs and special services – as well as the standard behaviour of the corresponding action. This will mean that the pages of Figs. 3.6 and 3.7 describe the standard behaviour of the telephone system, while the additional complexity introduced by time-outs and the special services are confined to the new pages. There will be a consistent relationship between the two levels of description – but this relationship will not be a strict behavioural equivalence.

The telephone system was described in a **top-down** manner, but this does not necessarily mean that it was constructed in this way. It could just as well have been constructed **bottom-up** or (more likely) by mixing the two strategies. When a page during a design gets too many details we can create a new subpage and move some of the details to this page We can construct pages modelling the low-level features of a system and then later add pages which describe the more high-level features (using the low-level building blocks via substitution transitions). The CPN editor described in Chap. 6 supports the creation of hierarchical nets, and it is very easy to add new subpages – or rearrange the page hierarchy in other ways. As an example, it would in a few minutes be possible to transform the hierarchical CP-net of Figs. 3.6–3.8 into the non-hierarchical CP-net of Fig. 3.9 (or vice versa). Finally, it should be remarked that each individual page of Figs. 3.6 and 3.7 contains fewer nodes than in most hierarchical CP-nets used in practice.

3.3 Structure of Hierarchical CP-nets

As described in Sects. 3.1 and 3.2, a hierarchical CP-net consists of a set of pages. Unfortunately "page" and "place" both start with the same letter, and thus we use S to denote the set of pages – it may help to think of the pages as subnets. Each page $s \in S$ is a non-hierarchical CP-net, i.e., a tuple:

$$(\Sigma_S, P_S, T_S, A_S, N_S, C_S, G_S, E_S, I_S).$$

To talk about the elements of the entire hierarchical CP-net, we use the notation:

$$\Sigma = \bigcup_{s \in S} \Sigma_S \qquad P = \bigcup_{s \in S} P_S \qquad T = \bigcup_{s \in S} T_S \qquad A = \bigcup_{s \in S} A_S$$

where it should be noted that the sets of colour sets usually have common elements, while the sets of net elements are required to be disjoint. As examples, the latter means that we cannot have a place which belongs to two different pages (or a place which is an arc of another page).

The disjointness of the net elements means that without ambiguity we can use a "global" colour set function $C \in [P \rightarrow \Sigma]$ instead of the "local" colour set functions $C_s \in [P_s \to \Sigma_s]$. The global function is defined from the local functions in the following way:

$$\forall s \in S\ \forall p \in P_s: C(p) = C_s(p).$$

Analogously, we can define global versions of the node function N, the guard function G, the arc expression function E, and the initialization function I. It also means that we can use p(a), t(a), s(a), d(a), $A(x_1,x_2)$, A(x), In(x), Out(x), X(x) and $E(x_1,x_2)$ – in exactly the same way as for non-hierarchical nets. When x_1 and x_2 belong to different pages $A(x_1,x_2) = \emptyset$ and $E(x_1,x_2) = \emptyset$.

The notational conventions described above allow us to shift our attention from a given page to the entire hierarchical CP-net by omitting the page index. However, it is sometimes also desirable to be able to do the opposite. As an example, from the set of all substitution nodes SN and the set of all port nodes PN, we define the set of substitution/port nodes belonging to a single page $s \in S$:

$$SN_s = SN \cap T_s \quad PN_s = PN \cap P_s.$$

As mentioned in Sect. 3.1, a place p is said to be a **socket node** of a substitution transition t iff $p \in X(t)$. Moreover, we define the **socket type** function, which maps from pairs of socket nodes and substitution transitions into {in, out, i/o}. The •t and t• notation was defined at the end of Sect. 2.2.

$$ST(p,t) = \begin{cases} \text{in} & \text{if } p \in (\bullet t - t\bullet) \\ \text{out} & \text{if } p \in (t\bullet - \bullet t) \\ \text{i/o} & \text{if } p \in (\bullet t \cap t\bullet). \end{cases}$$

Notice that p may simultaneously be a socket node of two (or more) transitions t_1 and t_2, and we may then have $ST(p,t_1) \neq ST(p,t_2)$. Now we are ready to define hierarchical CP-nets:

Definition 3.1: A **hierarchical CP-net** is a tuple HCPN = (S, SN, SA, PN, PT, PA, FS, FT, PP) satisfying the requirements below:

(i) S is a finite set of **pages** such that:
- Each page $s \in S$ is a non-hierarchical CP-net: $(\Sigma_s, P_s, T_s, A_s, N_s, C_s, G_s, E_s, I_s)$.
- The sets of net elements are pairwise disjoint: $\forall s_1,s_2 \in S: [s_1 \neq s_2 \Rightarrow (P_{s_1} \cup T_{s_1} \cup A_{s_1}) \cap (P_{s_2} \cup T_{s_2} \cup A_{s_2}) = \emptyset]$.

(ii) $SN \subseteq T$ is a set of **substitution nodes**.

(iii) SA is a **page assignment** function. It is defined from SN into S such that:
- No page is a subpage of itself: $\{s_0 s_1 \ldots s_n \in S^* \mid n \in \mathbb{N}_+ \wedge s_0 = s_n \wedge \forall k \in 1..n: s_k \in SA(SN_{s_{k-1}})\} = \emptyset$.

(iv) $PN \subseteq P$ is a set of **port nodes**.

(v) PT is a **port type** function. It is defined from PN into {in, out, i/o, general}.

(continues)

(vi) PA is a **port assignment** function. It is defined from SN into binary relations such that:

- Socket nodes are related to port nodes:
$\forall t \in SN$: $PA(t) \subseteq X(t) \times PN_{SA(t)}$.
- Socket nodes are of the correct type:
$\forall t \in SN\ \forall (p_1,p_2) \in PA(t)$: $[PT(p_2) \neq \text{general} \Rightarrow ST(p_1,t) = PT(p_2)]$.
- Related nodes have identical colour sets and equivalent initialization expressions:
$\forall t \in SN\ \forall (p_1,p_2) \in PA(t)$: $[C(p_1) = C(p_2) \wedge I(p_1)\langle\rangle = I(p_2)\langle\rangle]$.

(vii) $FS \subseteq P_S$ is a finite set of **fusion sets** such that:

- Members of a fusion set have identical colour sets and equivalent initialization expressions:
$\forall fs \in FS$: $\forall p_1,p_2 \in fs$: $[C(p_1) = C(p_2) \wedge I(p_1)\langle\rangle = I(p_2)\langle\rangle]$.

(viii) FT is a **fusion type** function. It is defined from fusion sets into {global, page, instance} such that:

- Page and instance fusion sets belong to a single page:
$\forall fs \in FS$: $[FT(fs) \neq \text{global} \Rightarrow \exists s \in S$: $fs \subseteq P_s]$.

(ix) $PP \in S_{MS}$ is a multi-set of **prime pages**.

(i) Each **page** is a non-hierarchical CP-net, and we require that none of these have any net elements (i.e., places, transitions and arcs) in common.

(ii) Each **substitution node** is a transition.

(iii) The **page assignment** relates substitution transitions to their subpages. We require that no page be a subpage of itself. Otherwise, the process of substituting supernodes with their direct subpages would be infinite and it would be impossible to construct an equivalent non-hierarchical CP-net (without allowing P, T and A to be infinite).

(iv) Each **port node** is a place. Notice that we allow a page to have port nodes even when it is not a subpage. Such port nodes have no semantic meaning (and thus they can be turned into non-ports without changing the behaviour of the CP-net).

(v) The **port type** divides the set of port nodes into input, output, input/output and general ports.

(vi) The **port assignment** relates socket nodes (i.e., the places surrounding a substitution transition) with port nodes (on the corresponding direct subpage). Each related pair of socket/port nodes must have matching socket/port types ("general" matches everything). Moreover, they must have identical colour sets and equivalent initialization expressions. We do not require the initialization expressions to be identical, but we do require that they (for the empty binding) evaluate to the same value. Notice that it is possible to relate several sockets to the same port, and vice versa. It is also possible to have sockets and ports which are totally unrelated. Many port assignments are bijective functions, defined on the set of all socket nodes. In this case there is a one-to-one correspondence between sockets and ports.

(vii) The **fusion sets** are subsets of P. Each of them may be empty and they are not demanded to be disjoint. All members of a fusion set must have identical colour sets and equivalent initialization expressions. Usually, it is only a few places that belong to fusion sets.

(viii) The **fusion type** divides the set of fusion sets into global, page and instance fusion sets. For the last two kinds of fusion sets all members must belong to the same page.

(ix) The **prime pages** is a multi-set over the set of all pages and they determine – together with the page assignment – how many instances the individual pages have. Often the prime page multi-set contains only a single page (with coefficient one).

To illustrate our formal definition of hierarchical CP-nets, Fig. 3.10 shows how the hierarchical CP-net in Figs. 3.6–3.8 can be represented as a many-tuple. To save space (and time) we do not give the tuple-definition of the individual pages and we abbreviate the page and node names to two or three letters (Cd ≈ Connected, while Cn ≈ Connection). It will be the only time in this book

(i) S = {Ph#1, EsC#2, Dia#3, BrE#4, BrS#5, BrR#6}.

(ii) SN = {EsC@Ph#1, Dia@EsC#2, BrE@EsC#2, BrS@Ph#1, BrR@Ph#1}.

(iii) SA(t) =
- EsC#2 if t=EsC@Ph#1
- Dia#3, if t=Dia@EsC#2
- BrE#4 if t=BrE@EsC#2
- BrS#5 if t=BrS@Ph#1
- BrR#6 if t=BrR@Ph#1.

(iv) PN = {In@EsC#2, Cd@EsC#2, Cn@EsC#2, In@Dia#3, No@Dia#3, Req@Dia#3, In@BrE#4, No@BrE#4, Req@BrE#4, In@BrS#5, Cd@BrS#5, Cn@BrS#5, In@BrR#6, Cd@BrR#6, Cn@BrR#6}.

(v) PT(p) =
- in if p∈ {In@Dia#3, No@BrE#4, Req@BrE#4, Cd@BrS#5, Cn@BrS#5, Cd@BrR#6, Cn@BrR#6}
- out if p∈ {Cd@EsC#2, Cn@EsC#2, No@Dia#3, Req@Dia#3, In@BrE#4, In@BrS#5, In@BrR#6}
- i/o if p∈ {In@EsC#2}.

(vi) PA(t) =
- {(In@Ph#1, In@EsC#2), (Cd@Ph#1, Cd@EsC#2), (Cn@Ph#1, Cn@EsC#2)} if t=EsC@Ph#1
- {(In@EsC#2, In@Dia#3), (No@EsC#2, No@Dia#3), (Req@EsC#2, Req@Dia#3)} if t=Dia@EsC#2
- ....

(vii) FS = {{En@EsC#2, En@Dia#3, En@BrE#4, En@BrS#5, En@BrR#6 }}.

(viii) FT(fs) = global for all fs∈FS.

(ix) PP = 1`Ph#1.

Fig. 3.10. The hierarchical CP-net from Figs. 3.6–3.8 represented as a many-tuple

we present a particular hierarchical CP-net as a tuple. Hopefully, Fig. 3.10 demonstrates that, in practice, it would be intolerable to work with hierarchical CP-nets without having a graphical representation of them.

Page instances

As described in Sect. 3.1, a page $s \in S$ may have many different page instances. First of all, s may itself be a member of the multi-set of the prime pages, and secondly it may be a subpage of such a page. In Def. 3.2 we use s* and n* to identify the element in PP from which the page instance is constructed, while $t_1 t_2 \dots t_m$ identifies a sequence of substitution transitions leading from s* to s. In this sequence each node t_k belongs to the direct subpage of its predecessor t_{k-1}. Notice that the sequence may be empty – indicating that the corresponding page instance exists because s itself belongs to PP.

Now we are ready to give a formal definition of page instances that uses the ideas in the intuitive explanation given above:

Definition 3.2: The set of **page instances** of a page $s \in S$ is the set SI_s of all triples $(s^*, n^*, t_1 t_2 \dots t_m)$ that satisfy the following requirements:

(i) $s^* \in PP \wedge n^* \in 1..PP(s^*)$.

(ii) $t_1 t_2 \dots t_m$ is a sequence of substitution nodes, with $m \in \mathbb{N}$, such that:
- $m = 0 \Rightarrow s^* = s$.
- $m > 0 \Rightarrow (t_1 \in SN_{s^*} \wedge [k \in 2..m \Rightarrow t_k \in SN_{SA(t_{k-1})}] \wedge SA(t_m) = s)$.

Page instances where the third component is the empty sequence are said to be **prime** page instances, while all others are **secondary** page instances.

To illustrate Def. 3.2 we give the page instances of the ring network from Sect. 3.1 and the telephone system from Sect. 3.2. However, it should be understood that we usually find the page instances by looking at the different routes in the page hierarchy graph, as explained in Sect. 3.1. The only purpose of Def. 3.2 is to give a precise and unambiguous definition of the set of such routes. This means that the examples below will be the only time in this book where we represent the page instances of a particular hierarchical net as triples. We use ε to denote the empty sequence, and we abbreviate some of the page/transition names:

Ring Network:

$SI_{NetW\#10}$ = {(NetW#10, 1, ε)}.

$SI_{Site\#11}$ = {(NetW#10, 1, Site1@NetW#10), (NetW#10, 1, Site2@NetW#10), (NetW#10, 1, Site3@NetW#10), (NetW#10, 1, Site4@NetW#10)}.

Phone System:

$SI_{Ph\#1}$ = {(Ph#1, 1, ε)}.

$SI_{EsC\#2}$ = {(Ph#1, 1, EsC@Ph#1)}.

$SI_{Dia\#3}$ = {(Ph#1, 1, EsC@Ph#1 Dia@EsC#2)}.

$SI_{BrE\#4}$ = {(Ph#1, 1, EsC@Ph#1 BrE@EsC#2)}.

$SI_{BrS\#5}$ = {(Ph#1, 1, BrS@Ph#1)}.

$SI_{BrR\#6}$ = {(Ph#1, 1, BrR@Ph#1)}.

It can be seen that each page has a single instance – except for *Site#11* which has four. The instances of *NetWork#10* and *Phone#1* are primary page instances. This means that they exist because the corresponding pages belong to the multi-set of prime pages. All other page instances are secondary page instances. This means that they exist because the corresponding pages are subpages of other pages (which belong to the multi-set of prime pages).

All the page instances in the above examples have 1 as their second component. This will always be the case when the multi-set of prime pages is a set, i.e., when $0 \leq PP(s) \leq 1$ for all pages $s \in S$. If we instead define the multi-set of prime pages of the ring network to be 4\`Site#11, we get the following page instances:

Ring Network:

$SI_{NetW\#10} = \emptyset$.

$SI_{Site\#11} = \{(Site\#11, 1, \varepsilon), (Site\#11, 2, \varepsilon), (Site\#11, 3, \varepsilon), (Site\#11, 4, \varepsilon)\}$.

Place, transition and arc instances

When a page has several page instances, these each have their own instances of the corresponding places, transitions and arcs. However, it should be noted that substitution nodes and their surrounding arcs do not have instances – because, intuitively, they are replaced by instances of the corresponding direct subpages:

Definition 3.3: The set of **place instances** of a page $s \in S$ is the set PI_S of all pairs (p,id) that satisfy the following requirements:

(i) $p \in P_S$.
(ii) $id \in SI_S$.

The set of **transition instances** of a page $s \in S$ is the set TI_S of all pairs (t,id) that satisfy the following requirements:

(iii) $t \in T_S - SN_S$.
(iv) $id \in SI_S$.

The set of **arc instances** of a page $s \in S$ is the set AI_S of all pairs (a,id) that satisfy the following requirements:

(v) $a \in A_S - A(SN_S)$.
(vi) $id \in SI_S$.

Each place instance, transition instance and arc instance is said to **belong** to the page instance in its second component.

Above we have defined the place instances, transition instances and arc instances in a hierarchical CP-net. However, we have to remember that some of the place instances are related to each other – because of the fusion sets and because of the port assignments. Two place instances (p_1,id_1) and (p_2,id_2) are related by a fusion set $fs \in FS$ iff the following conditions are fulfilled:

- The two original places must both belong to fs. This means that $p_1, p_2 \in fs$.
- When fs is an instance fusion set, the two place instances must belong to the same page instance. This means that $id_1 = id_2$.
- When fs is a global fusion set or a page fusion set, there is no restriction on the relation between id_1 and id_2.

Analogously, two place instances (p_1, id_1) and (p_2, id_2) are related by the port assignment of a substitution transition $t \in SN$ iff the following conditions are fulfilled:

- The two original places must be related by the port assignment. This means that $(p_1, p_2) \in PA(t)$.
- The page instance $id_2 = (s_2, n_2, tt_2)$ of the port node p_2 must be a secondary page instance created because of the existence of the substitution transition t on the page instance $id_1 = (s_1, n_1, tt_1)$ of the socket node p_1. This means that the two page instances must originate from the same prime page instance, i.e., that $(s_1, n_1) = (s_2, n_2)$. Moreover, id_2 must have the same sequence of substitution nodes as id_1, except that t has been added. This means that $tt_1 \hat{} t = tt_2$, where we use ^ to denote concatenation of sequences.

Above we have described how two place instances – due to a fusion set or a port assignment – may become related to each other. However, we want the page instance relation to be transitive, symmetric, and reflexive – and thus we extend the above relation to an equivalence relation. This means that we get the following definition, where PI denotes the set of all place instances of the entire CP-net (i.e., the union of PI_s taken over $s \in S$):

Definition 3.4: The **place instance relation** is the smallest equivalence relation on PI containing all those pairs $((p_1,(s_1,n_1,tt_1)), (p_2,(s_2,n_2,tt_2))) \in PI \times PI$ that satisfy at least one of the following conditions:

(i) $\exists fs \in FS: [p_1, p_2 \in fs \ \wedge \ [FT(fs) = \text{instance} \Rightarrow (s_1,n_1,tt_1) = (s_2,n_2,tt_2)]]$.

(ii) $\exists t \in SN: [(p_1,p_2) \in PA(t) \ \wedge \ (s_1,n_1) = (s_2,n_2) \ \wedge \ tt_1 \hat{} t = tt_2]$.

An equivalence class of the place instance relation is called a **place instance group** and the set of all such equivalence classes is denoted by PIG.

To illustrate Def. 3.4 we show those place instance groups of the ring network from Sect. 3.1 which have more than one element when both the port assignment and the two fusion sets are taken into account. However, it should be understood that we usually do not make an explicit construction of the place instance groups. Instead we find the relationship between place instances by looking at the port assignments and the fusion sets, as explained in Sect. 3.1. The only purpose of Def. 3.4 is to give a precise and unambiguous definition of the interpretation of the port assignments and the fusion sets. This means that the examples below will be the only time in this book where we explicitly construct the place instance groups of a particular hierarchical net:

Due to port assignments:

{(1to2@NetW#10, (NetW#10, 1, ε)), (Outg@Site#11, (NetW#10, 1, Site1)), (Inc@Site#11, (NetW#10, 1, Site2))}.

{(2to3@NetW#10, (NetW#10, 1, ε)), (Outg@Site#11, (NetW#10, 1, Site2)), (Inc@Site#11, (NetW#10, 1, Site3))}.

{(3to4@NetW#10, (NetW#10, 1, ε)), (Outg@Site#11, (NetW#10, 1, Site3)), (Inc@Site#11, (NetW#10, 1, Site4))}.

{(4to1@NetW#10, (NetW#10, 1, ε)), (Outg@Site#11, (NetW#10, 1, Site4)), (Inc@Site#11, (NetW#10, 1, Site1))}.

Due to the instance fusion set:

{(RecI@Site#11, (NetW#10, 1, Site1)), (RecE@Site#11, (NetW#10, 1, Site1))}.
{(RecI@Site#11, (NetW#10, 1, Site2)), (RecE@Site#11, (NetW#10, 1, Site2))}.
{(RecI@Site#11, (NetW#10, 1, Site3)), (RecE@Site#11, (NetW#10, 1, Site3))}.
{(RecI@Site#11, (NetW#10, 1, Site4)), (RecE@Site#11, (NetW#10, 1, Site4))}.

Due to the page fusion set:

{(SentE@Site#11, (NetW#10, 1, Site1)), (SentE@Site#11, (NetW#10, 1, Site2)), (SentE@Site#11, (NetW#10, 1, Site3)), (SentE@Site#11, (NetW#10, 1, Site4))}.

When a hierarchical CP-net is translated into its non-hierarchical equivalent (to be defined in Sect. 3.5), the latter contains a place for each place instance group of the former. In Fig. 3.5, the four places *1to2, 2to3, 3to4,* and *4to1* correspond to the first four place instance groups above (those due to port assignments). Analogously, the four places *RecExt/Int1, RecExt/Int2, RecExt/Int3,* and *RecExt/Int4* correspond to the next four place instance groups (those due to the instance fusion set), while the place *SentExt* corresponds to the place instance group due to the page fusion set. Each of the remaining places in Fig. 3.5 corresponds to a place instance group with only one element.

3.4 Behaviour of Hierarchical CP-nets

Having defined the static structure of hierarchical CP-nets we are now ready to consider their behaviour. We use p and t to denote a place and a transition, p' and t' to denote a place instance and a transition instance, and p" to denote a place instance group. First we extend the arc expression function E so that it can be applied to place instances and transition instances (instead of just places and transitions). The intuition behind the extension is as follows. We want a place instance $p'=(p,id_p)$ on a page instance id_p to be related to a transition instance $t'=(t,id_t)$ on the *same* page instance (i.e., with $id_p = id_t$) in exactly the same way that the original place p was related to the original transition t. This means that $E(p',t') = E(p,t)$ and $E(t',p') = E(t,p)$. In contrast to this, we want a pair of instances $p'=(p,id_p)$ and $t' = (t,id_t)$ which belong to *different* page instances (i.e., with $id_p \neq id_t$) to be unrelated. This means that $E(p',t') = E(t',p') = \emptyset$. Intuitively this captures the fact that each page instance is a copy of the corresponding page (with substitution transitions and their surrounding arcs removed). The page instance has the same arc expressions as the original page and it has no arcs to/from other page instances:

- $\forall p'=(p,id_p)\in PI\ \forall t'=(t,id_t)\in TI$:

$$[\ id_p = id_t \Rightarrow (E(p',t') = E(p,t) \wedge E(t',p') = E(t,p)) \wedge$$
$$id_p \neq id_t \Rightarrow (E(p',t') = E(t',p') = \emptyset)\ ].$$

Next we observe that the definition of place instance groups, together with Def. 3.1 (vi)+(vii), guarantee that all members of a place instance group originate from places which have the same colour set and equivalent initialization expressions (i.e., initialization expressions which evaluate to the same multi-set). Thus we can define token elements, binding elements, markings and steps as shown below – where it should be noted that (because of the remarks above) it does not matter which representative we choose for a place instance group. Intuitively, the definitions are analogous to those of a non-hierarchical CP-net – except that we replace places by place instance groups (i.e., P by PIG), while we replace transitions by transition instances (i.e., T by TI). Following the standard notation for equivalence classes we use [p'] and $[(p,id_p)]$ to denote the place instance group to which a given place instance $p'=(p,id_p)$ belongs:

Definition 3.5: A **token element** is a pair (p'',c) where $p''=[(p,id_p)]\in PIG$ and $c\in C(p)$, while a **binding element** is a pair (t',b) where $t'=(t,id_t)\in TI$ and $b\in B(t)$. The set of all token elements is denoted by TE while the set of all binding elements is denoted by BE.

A **marking** is a multi-set over TE while a **step** is a *non-empty* and *finite* multi-set over BE. The **initial marking** M_0 is the marking which is obtained by evaluating the initialization expressions:

$$\forall(p'',c)=([(p,id)],c)\in TE\text{: } M_0(p'',c) = (I(p))(c).$$

The sets of all markings and steps are denoted by $\mathbb{M}$ and $\mathbb{Y}$, respectively.

Analogously to non-hierarchical nets, notice that there is a unique correspondence between markings and functions M* defined on PIG such that $M^*([(p,id)])\in C(p)_{MS}$ for all $[(p,id)]\in PIG$, and a unique correspondence between steps Y and functions Y* defined on TI such that $Y^*(t,id)\in B(t)_{MS}$ is finite for all $(t,id)\in TI$ and non-empty for at least one $(t,id)\in TI$. Thus we shall often represent markings as functions defined on PIG and steps as functions defined on TI. For brevity we shall also use M(p,id) as a shorthand for M([(p,id)]). Intuitively this means that we consider each place instance to have a marking which is identical to the marking of its place instance group.

Now we are ready to give the formal definition of enabling. The definition is analogous to the definition for non-hierarchical CP-nets, except that in the sums we now have to consider not only all the binding elements (t', b) in the step Y, but also all the place instances p' in the place instance group p'':

Definition 3.6: A step Y is **enabled** in a marking M iff the following property is satisfied:

$$\forall p'' \in PIG: \sum_{\substack{(t',b) \in Y \\ p' \in p''}} E(p',t')\langle b\rangle \le M(p'').$$

We define **enabled** transition instances and **concurrently enabled** transition instances/binding elements analogously to the corresponding concepts in a non-hierarchical CP-net.

When a step is enabled in a marking M_1 it may **occur**, changing the marking M_1 to another marking M_2, defined by:

$$\forall p'' \in PIG: M_2(p'') = \Big(M_1(p'') - \sum_{\substack{(t',b) \in Y \\ p' \in p''}} E(p',t')\langle b\rangle\Big) + \sum_{\substack{(t',b) \in Y \\ p' \in p''}} E(t',p')\langle b\rangle.$$

We define **removed/added tokens**, **direct reachability**, **occurrence sequences** and **reachability** analogously to the corresponding concepts for non-hierarchical CP-nets.

Let us finish this section by discussing how our formal definition of hierarchical CP-nets can be modified to allow arc expressions, guards and initialization expressions which depend upon the page instance – via the predeclared page instance function inst() (cf. the ring network in Sect. 3.1). However, this is rather straightforward: first of all we must give each page instance dependent expression an extra variable, which always is bound to the page instance number. Secondly we must weaken the constraints in Def. 3.1 (vi)+(vii) so that they only require the initialization expressions to be equivalent for those place instances that belong to the same place instance group. Finally, in Sect. 3.5 – when we construct the non-hierarchical equivalent – we must replace all appearances of the page instance function inst() with the corresponding page instance number.

3.5 Equivalent Non-Hierarchical CP-nets

This section is theoretical and it can be skipped by readers who are primarily interested in the practical application of CP-nets. The section investigates the formal relationship between hierarchical CP-nets and non-hierarchical CP-nets. For each hierarchical CP-net we show how to construct an equivalent non-hierarchical CP-net. The existence of the non-hierarchical equivalent is extremely useful, because it tells us how to generalize the basic concepts and analysis methods of non-hierarchical CP-nets to hierarchical CP-nets. We simply define these concepts in such a way that a hierarchical CP-net has a given property iff the non-hierarchical equivalent has the corresponding property. It is important to understand that we never make the translation for a particular hierarchical CP-net. When we describe a system we use hierarchical CP-nets directly, without constructing the non-hierarchical equivalent. Analogously, we analyse a hierarchical CP-net directly, without constructing the non-hierarchical equivalent. The intuition behind the construction of the non-hierarchical equivalent is as follows:

- First we make as many copies of each page as it has page instances in the hierarchical net. We position all these copies next to each other on a huge table, or on a soccer field.
- Then we remove all substitution transitions, together with their surrounding arcs. Now we have a number of disconnected nets, which together have a place for each place instance of the hierarchical net, a transition for each transition instance, and an arc for each arc instance.
- Finally we merge all the members of each place instance group into a single place which has all the arcs of the original place instances. Now we have the hierarchical equivalent. It has a place for each place instance group of the hierarchical net, a transition for each transition instance, and an arc for each arc instance.

Now we give the formal definition of the non-hierarchical equivalent. It should of course be verified that the defined tuple really is a non-hierarchical CP-net, i.e., that it satisfies the constraints in Def. 2.5. However, this verification is straightforward and thus it is omitted:

Definition 3.7: Let a hierarchical CP-net HCPN = (S, SN, SA, PN, PT, PA, FS, FT, PP) be given. Then we define the **equivalent non-hierarchical CP-net** to be CPN* = (Σ*, P*, T*, A*, N*, C*, G*, E*, I*) where:

(i) $\Sigma^* = \Sigma$.
(ii) $P^* = PIG$.
(iii) $T^* = TI$.
(iv) $A^* = AI$.
(v) $\forall a^*=(a,id)\in A^*\ \forall (p,t)\in P\times T$:
$[N(a) = (p,t) \Rightarrow N^*(a^*) = ([(p,id)],(t,id)) \wedge$
$N(a) = (t,p) \Rightarrow N^*(a^*) = ((t,id),[(p,id)])]$.
(vi) $\forall p^*=[(p,id)]\in P^*$: $C^*(p^*) = C(p)$.
(vii) $\forall t^*=(t,id)\in T^*$: $G^*(t^*) = G(t)$.
(viii) $\forall a^*=(a,id)\in A^*$: $E^*(a^*) = E(a)$.
(ix) $\forall p^*=[(p,id)]\in P^*$: $I^*(p^*) = I(p)$.

(i) The non-hierarchical CP-net has the same set of colour sets as the hierarchical CP-net.

(ii) The non-hierarchical CP-net has a place for each place instance group of the hierarchical CP-net. This means that there is a place for each place instance – unless that place instance either belongs to a fusion set (in which case the place instance is merged with instances of other members of the fusion set) or is an assigned socket/port node (in which case it is merged with the place instance to which it is assigned).

(iii) + (iv) The non-hierarchical CP-net has a transition for each transition instance of the hierarchical CP-net. Analogously, it has an arc for each arc instance of the hierarchical CP-net.

(v) The basic idea behind the definition of the node function is that each page instance has the same arcs as the original page (except that we omit the arcs surrounding substitution nodes). This means that a place instance and a transition instance can have connecting arcs only if they belong to the same page instance – and in that case they have connecting arcs iff the original place and transition have. However, it should be noted that (due to place fusion and socket/port assignment) the node function maps into place instance groups (and not into individual place instances). This is done in such a way that each place instance group gets a set of surrounding arcs that is the union of those arcs that the corresponding place instances would have got (if they had not participated in any fusion or socket/port assignment).

(vi) The colour set of a place instance group is determined by the colour set of one of the participating places. From Def. 3.1 (vi)+(vii) it follows that all these places must have identical colour sets.

(vii) The guard of a transition instance is determined by the guard of the corresponding transition.

(viii) The arc expression of an arc instance is determined by the arc expression of the corresponding arc.

(ix) The initialization expression of a place instance group is determined by the initialization expression of one of the participating places. From Def. 3.1 (vi)+(vii) it follows that all these places must have initialization expressions which evaluate to the same value, and thus it does not matter which one we choose.

The following theorem shows that each hierarchical CP-net has exactly the same sets of markings, steps and occurrence sequences as the non-hierarchical equivalent, and thus the two nets are behaviourally equivalent. All concepts with a star refer to CPN*, while those without refer to HCPN:

Theorem 3.8: Let HCPN be a hierarchical CP-net and let CPN* be the non-hierarchical equivalent. Then we have the following properties:

(i) $\mathbb{M} = \mathbb{M}^* \;\wedge\; M_0 = M_0^*$.
(ii) $\mathbb{Y} = \mathbb{Y}^*$.
(iii) $\forall M_1, M_2 \in \mathbb{M}\ \forall Y \in \mathbb{Y}: M_1\,[Y\rangle_{HCPN}\, M_2 \Leftrightarrow M_1\,[Y\rangle_{CPN^*}\, M_2$.

Proof: The proof is a simple consequence of our earlier definitions.

Property (i): From Def. 2.7 we have $\mathbb{M}^* = TE^*_{MS}$ where TE* consists of all pairs (p*,c) with $p^* \in P^*$ and $c \in C^*(p^*)$. From Def. 3.5 we have $\mathbb{M} = TE_{MS}$ where TE consists of all pairs (p",c) with $p" = [(p, id_p)] \in PIG$ and $c \in C(p)$. Thus it is sufficient to prove that P* = PIG and C*(p*) = C(p), but this follows from Def. 3.7 (ii) and (vi).

Next let us prove that the two initial markings are identical. From Def. 3.5 we have:

(*) $\forall (p",c) = ([(p, id_p)], c) \in TE: M_0(p",c) = (I(p))(c)$.

From Def. 2.7 we have:

$$\forall (p^*,c) \in TE^*: M_0{}^*(p^*,c) = (I^*(p^*))(c),$$

which by Def. 3.7 (ii) and Def. 3.5 is equivalent to:

$$\forall (p'',c)=([(p,id_p)],c) \in TE: M_0{}^*(p'',c) = (I^*(p''))(c),$$

which by Def. 3.7 (ix) is equivalent to:

$$\forall (p'',c)=([(p,id_p)],c) \in TE: M_0{}^*(p'',c) = (I(p))(c),$$

which has the same form as (*).

Property (ii): From Def. 2.7 we have that $\mathbb{Y}^*$ consists of all non-empty and finite multi-sets in BE^*_{MS} where BE^* consists of all pairs (t^*,b) with $t^* \in T^*$ and $b \in B^*(t^*)$. From Def. 3.5 we have that $\mathbb{Y}$ consists of all non-empty and finite multi-sets in BE_{MS} where BE consists of all pairs (t',b) with $t'=(t,id_t) \in TI$ and $b \in B(t)$. Thus it is sufficient to prove that $T^* = TI$ and $B^*(t^*) = B(t)$, but this follows from Def. 3.7 (iii), (iv), (v), (vii) and (viii).

Property (iii): First we prove that the enabling rules coincide, i.e., that:

$$M_1\, [Y\rangle_{HCPN} \Leftrightarrow M_1\, [Y\rangle_{CPN^*}.$$

From Def 3.6 it follows that $M_1\, [Y\rangle_{HCPN}$ iff:

$$(**) \quad \forall p'' \in PIG: \sum_{\substack{(t',b) \in Y \\ p' \in p''}} E(p',t')<b> \;\le\; M_1(p'').$$

From Def. 2.8 it follows that $M_1\, [Y\rangle_{CPN^*}$ iff:

$$\forall p^* \in P^*: \sum_{(t^*,b) \in Y} E^*(p^*,t^*)<b> \;\le\; M_1(p^*),$$

which by Def. 3.7 (ii)+(iii) is equivalent to:

$$\forall p'' \in PIG: \sum_{(t',b) \in Y} E^*(p'',t')<b> \;\le\; M_1(p''),$$

which by Def. 3.7 (iv)+(v)+(viii)+(ix), and the extension of E^* from A^* to $P^* \times T^* \cup T^* \times P^*$, is equivalent to:

$$\forall p'' \in PIG: \sum_{\substack{(t',b) \in Y \\ p' \in p''}} E(p',t')<b> \;\le\; M_1(p''),$$

which is identical to (**).

Next we can prove that the occurrence rules coincide, i.e., that

$$M_1\, [Y\rangle_{HCPN} M_2 \Leftrightarrow M_1\, [Y\rangle_{CPN^*} M_2.$$

However, this is done in a way which is totally analogous to the proof above, and thus we shall omit it. □

It is also possible to go in the other direction and construct a hierarchical CP-net from a given non-hierarchical CP-net. The simplest way to do this is as follows:

- Let the hierarchical net have a single page s which is identical to the non-hierarchical net.
- Let the substitution nodes SN, the port nodes PN, and the fusion sets FS be empty. This means that the page assignment, port type, port assignment and fusion type functions have an empty domain and thus they become trivial.
- Let the multi-set of prime pages have one appearance of s.

Bibliographical Remarks

Hierarchical CP-nets were developed together with Peter Huber and Robert M. Shapiro. They were first presented in [49], which describes five different hierarchy constructs, known as substitution transitions, substitution places, invocation transitions, fusion places and fusion transitions. The semantics of the five constructs was only informally explained, by sketching the construction of the non-hierarchical equivalent.

The CPN tools described in Chap. 6 support substitution transitions and fusion places, and thus there have been many experiences with these two hierarchy constructs. The experiences have shown the two hierarchy constructs to be very useful, without revealing any theoretical problems. Formal definitions of the structure and behaviour of hierarchical CP-nets, with substitution transitions and fusion places, were first presented in [58]. These definitions are analogous to the definitions in Sects. 3.3–3.5 of this book.

Experiments with the other three hierarchy constructs – substitution places, invocation transitions, and fusion transitions – have disclosed some practical and theoretical problems. Simultaneously, it has turned out that it is possible to merge many of the basic ideas behind these hierarchy constructs into a single new construct.

In Chap. 7 we describe a number of industrial applications of hierarchical CP-nets – more information about some of these can be found in [50], [87], [100] and [101]. Proposals of different sets of hierarchy constructs can be found in [33] and [63].

Exercises

Exercise 3.1.
Consider the non-hierarchical CP-net used to model a telephone system in Exercise 1.8 (a).

(a) Change the net to a hierarchical CP-net. Make a simulation of the constructed net.

(b) Compare your hierarchical net to the one in Sect. 3.2.

Exercise 3.2.
Consider the hierarchical telephone system from Sect. 3.2.

(a) Convince yourself that the many-tuple in Fig. 3.10 fulfils the requirements in Def. 3.1.
(b) Find all those place instance groups which have more than one element.
(c) Convince yourself that Fig. 3.9 is the non-hierarchical equivalent of Figs. 3.6–3.8.

Exercise 3.3.
Consider a production system with the following page hierarchy graph:

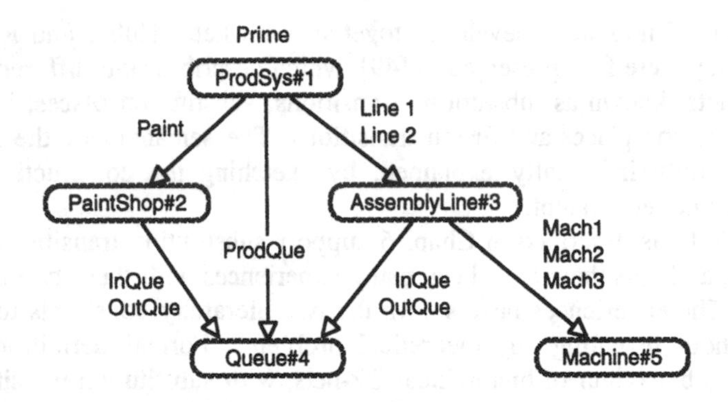

(a) Find the number of page instances for each page.
(b) Represent some typical page instances as triples (i.e., in the form used in Def. 3.2).

Exercise 3.4.
Consider the data base system from Sect. 1.3.

(a) Modify the CP-net in Fig. 1.13 so that it becomes a hierarchical CP-net with one prime page and two subpages. One of the subpages describes the activities of the sender (i.e., the left-hand side of Fig. 1.13), while the other describes the activities of the receivers (i.e., the right-hand side of Fig. 1.13). Make a simulation of the hierarchical CP-net.
(b) Modify the CP-net in Fig. 1.13 so that it becomes a hierarchical CP-net with two prime pages and *no* subpages. The two pages are related to each other by means of three global fusion sets. One of the pages describes the activities of the sender, while the other describes the activities of the receivers. Make a simulation of the hierarchical CP-net.
(c) Compare the two hierarchical CP-nets from (a) and (b). Which one do you prefer?

Exercise 3.5.
Consider the master/slave system from Exercise 1.9.

(a) Modify your CP-net so that it becomes a hierarchical CP-net with one prime page and two subpages. One of the subpages describes the activities of the master, while the other describes the activities of the slaves. Make a simulation of the hierarchical CP-net.

Exercise 3.6.
The ring network from Sect. 3.1 can also be modelled by a non-hierarchical CP-net, where the net structure is similar to the net structure of *Site#11*. Then all colour sets have an extra component saying which site the token "belongs to".

(a) Construct the non-hierarchical CP-net described above, and make a simulation of it.

(b) Compare the non-hierarchical CP-net from (a) with the non-hierarchical CP-net in Fig. 3.3. Do the two CP-nets have the same equivalent PT-net?

Chapter 4

Dynamic and Static Properties of Coloured Petri Nets

Dynamic properties characterize the behaviour of individual CP-nets, e.g., whether it is possible to reach a marking in which no step is enabled. It is often rather difficult to verify dynamic properties – in particular when relying only on informal arguments. However, in Chap. 5 we shall introduce a number of formal analysis methods which can be used to prove dynamic properties. The details of these analysis methods will be the prime contents of Vol. 2.

Static properties can be decided from the definition of individual CP-nets without considering the possible occurrence sequences. The main purpose of static properties is to characterize subclasses of CP-nets with nice characteristics. The study of static properties is, however, one of the rather few areas in which the knowledge and experience from low-level nets have not yet been adequately transferred to high-level nets.

Dynamic properties are also called behavioural properties, while static properties are called structural properties.

Most of the properties which are introduced in this chapter are defined for hierarchical CP-nets. However, as mentioned at the end of Sect. 3.5, each non-hierarchical CP-net determines a hierarchical CP-net with only one page (and one page instance). Thus we can also use the concepts for non-hierarchical nets.

Section 4.1 defines subnets and different static properties such as uniform, conservative and state machine nets. The next four sections define different dynamic properties. Section 4.2 introduces boundedness properties, while Sect. 4.3 deals with home properties, Sect. 4.4 with liveness properties and Sect. 4.5 with fairness properties. All the sections illustrate the defined properties by means of some of the CP-nets from Chaps. 1–3.

4.1 Static Properties

First we introduce bipartite directed graphs. It should be noted that, in contrast to classical graph theory, we allow a bipartite directed graph to have several arcs between the same ordered pair of nodes (and thus we define A as a separate set and not as a subset of $V_1 \times V_2 \cup V_2 \times V_1$):

Definition 4.1: A **bipartite directed graph** is a tuple BDG = (V_1, V_2, A, N) satisfying the requirements below:

(i) V_1 is a finite set of **nodes**.
(ii) V_2 is a finite set of **nodes**.
(iii) A is a finite set of **arcs** such that:
- $V_1 \cap V_2 = V_1 \cap A = V_2 \cap A = \emptyset$.

(iv) N is a **node** function. It is defined from A into $V_1 \times V_2 \cup V_2 \times V_1$.

From Def. 2.5 and Def 4.1 it follows that each non-hierarchical CP-net CPN = (Σ, P, T, A, N, C, G, E, I) determines a bipartite directed graph BDG = (P, T, A, N), which we shall call the **net structure** of the CP-net. For a function F, with domain D, we use $F \mid S$ to denote the restriction of F to a subset $S \subseteq D$.

Definition 4.2: Let BDG = (V_1, V_2, A, N) and BDG* = (V_1^*, V_2^*, A*, N*) be two bipartite directed graphs. BDG* is a **subgraph** of BDG iff the following properties are satisfied:
(i) $V_1^* \subseteq V_1$.
(ii) $V_2^* \subseteq V_2$.
(iii) $A^* \subseteq \{a \in A \mid N(a) \in V_1^* \times V_2^* \cup V_2^* \times V_1^*\}$.
(iv) $N^* = N \mid A^*$.

When the inclusion sign of (iii) can be replaced by an equality sign, BDG* is an **induced subgraph** of BDG.

Next we define subnets. Intuitively, we get a subnet of a non-hierarchical CP-net when we replace the net structure by a subgraph of it, and restrict the domain of the colour, guard, arc expression and initialization functions to those net elements which are part of the new net structure.

Definition 4.3: Let CPN = (Σ, P, T, A, N, C, G, E, I) and CPN* = (Σ*, P*, T*, A*, N*, C*, G*, E*, I*) be two non-hierarchical CP-nets. CPN* is a **subnet** of CPN iff the following properties are satisfied:

(i) $\Sigma^* \subseteq \Sigma$.
(ii) $P^* \subseteq P$.
(iii) $T^* \subseteq T$.
(iv) $A^* \subseteq \{a \in A \mid N(a) \in P^* \times T^* \cup T^* \times P^*\}$.
(v) $N^* = N \mid A^*$. *(continues)*

(vi) $C^* = C \mid P^*$.
(vii) $G^* = G \mid T^*$.
(viii) $E^* = E \mid A^*$.
(ix) $I^* = I \mid P^*$.

When the inclusion signs of (i) and (iv) can be replaced by equality signs, CPN* is an **induced subnet** of CPN.

The **border nodes** of CPN* are those nodes which have an arc connecting them with a node outside CPN*, i.e., the members of the set:

$$\{p \in P^* \mid \exists a \in A(p): t(a) \notin T^*\} \cup \{t \in T^* \mid \exists a \in A(t): p(a) \notin P^*\}.$$

CPN* is an **open** subnet iff all border nodes are places, and it is a **closed** subnet iff all border nodes are transitions.

It should be noted that the places and transitions of an induced subnet uniquely determine the set of arcs and the rest of the subnet. Thus we can make the following definition – where X(P") and X(T") denote the surrounding nodes of P" and T", calculated with respect to the entire net (cf. Sect. 2.2):

Definition 4.4: Let $P'' \subseteq P$ and $T'' \subseteq T$ be given. The subnets **induced** by (P",T"), P" and T" are defined as follows:

(i) SubNet(P",T") is the induced subnet with places P" and transitions T".
(ii) SubNet(P") = SubNet(P",X(P")).
(iii) SubNet(T") = SubNet(X(T"),T").

Each induced subnet is induced by its own set of places together with its own set of transitions. A subnet induced by a set of places is closed, while a subnet induced by a set of transitions is open. However, there are closed subnets which cannot be induced by a set of places, and analogously there are open subnets which cannot be induced by a set of transitions. The entire net is an open and closed subnet (of itself). The terminology "open" and "closed" reflects the fact that it is possible to define a topology on the nodes of a Petri net in such a way that the open subnets correspond to the open sets of the topology, and the closed subnets correspond to the closed sets. The definition of such a topology is outside the scope of this book. A reference to it can be found in the bibliographical remarks.

It should be noted that the definitions of border nodes, closed/open subnets, and subnets induced by a set of places and/or transitions only refer to the net structure. This means that they can also be applied to bipartite directed graphs. The same is true for the notation s(a), d(a), $A(x_1,x_2)$, A(x), In(x), Out(x) and X(x), defined in Sect. 2.2. It should also be noted that the concepts in Defs. 4.3–4.4 only are defined for non-hierarchical CP-nets. However, they can be used for the individual pages of a hierarchical net – since each of these is a non-hierarchical net. This means that we can speak, e.g., about the net structure of a page and a subnet of a page. All the concepts in the remaining parts of this chapter will be defined for hierarchical CP-nets – but they can also be used for non-hierarchical nets (as explained in the introduction to this chapter). Then it does not make sense to distinguish between places and place instances, or between transitions and transition instances. Moreover, we can use token elements and binding elements as introduced in Def. 2.7 (instead of the more complex Def. 3.5).

Definition 4.5: Let an arc $a \in A$ with arc expression E(a), a transition $t \in T$ and a non-negative integer $n \in \mathbb{N}$ be given:

(i) E(a) is **uniform** with **multiplicity** n iff:
$\forall b \in B(t(a)): |E(a)\langle b\rangle| = n.$

(ii) t is **uniform** iff all the surrounding arcs have a uniform arc expression.

(iii) t is **conservative** iff:
$$\forall b \in B(t): \sum_{p \in In(t)} |E(p,t)\langle b\rangle| = \sum_{p \in Out(t)} |E(t,p)\langle b\rangle|.$$

(iv) t has the **state machine** property iff:
$$\forall b \in B(t): \sum_{p \in In(t)} |E(p,t)\langle b\rangle| = \sum_{p \in Out(t)} |E(t,p)\langle b\rangle| = 1.$$

We use $|E(a)|$ to denote the multiplicity of a uniform arc expression E(a).

It should be noticed that a transition can be uniform without being conservative and vice versa. A CP-net is uniform iff all transitions in T–SN, are uniform. Conservative and state machine CP-nets are defined analogously. Below we illustrate the definitions of this section by applying them to some of the CP-nets from Chaps. 1–3. To avoid too much page flipping we suggest that you make a photocopy of the corresponding nets, i.e., Fig. 1.7, Fig. 1.13, Figs. 3.1–3.2, Figs. 3.6–3.7 and Fig. 3.9.

Data base

The CP-net is uniform and conservative. Figure 4.1 contains two induced subnets of the data base system. For brevity we have omitted the declarations. The left subnet has two transitions as border nodes, and thus it is closed. It is induced by

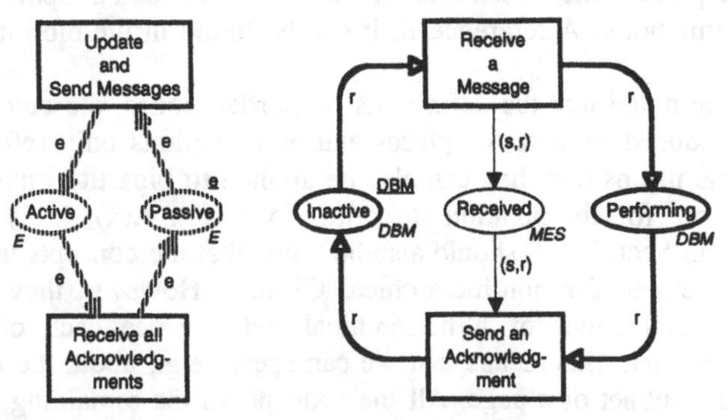

Fig. 4.1. Two induced subnets of the data base net in Fig. 1.13

the two places, and it is a state machine. The right subnet has two transitions and a place as border nodes, and thus it is neither open nor closed.

Resource allocation

Only the arc expressions (x,i), e and 2`e are uniform. They have multiplicity 1, 1 and 2, respectively. Only, the transitions T1 and T4 are uniform. SubNet ({A,B,C,D,E}) is a state machine. SubNet ({T1,T2,T3,T4,T5}) is the entire CP-net.

Telephone

The CP-net is uniform. *LiftRepl@BreakRec#6* is the only transition that has the state machine property, while *LiftRinging@EstabCon#2* and *RepConRec@BreakRec#6* are the only ones that are conservative. For the non-hierarchical equivalent in Fig. 3.9 two subnets are of particular interest. Both are state machines. One of the subnets describe the senders (and it consists of all those places, transitions and arcs which have a thick line). The other subnet describes the recipients (and it consists of all those places, transitions and arcs which have a shaded line). Nodes with both kinds of lines belong to both subnets. Neither of the two subnets is induced (e.g., because they do not contain all the arcs between *LiftRinging* and *Connected* – and because they do not contain all the (invisible) arcs that surround *Engaged).*

Ring network

The arc expressions n, n+1, {se=S(inst()), re=r, no=n} and p are the only ones that are uniform. They all have multiplicity 1. *NewPack* is the only transition that is uniform, while *Receive* is the only one that has the state machine property.

4.2 Boundedness Properties

Intuitively, upper and lower bounds tell us how many and how few token elements we can get, e.g., how many tokens we can have of a particular colour on a particular place instance. Notice that a multi-set is a function and thus it makes good sense to restrict it to a subset of its domain:

Definition 4.6: Let a set of token elements $X \subseteq TE$ and a non-negative integer $n \in \mathbb{N}$ be given.

(i) n is an **upper bound** for X iff:
$\forall M \in [M_0\rangle: |(M|X)| \leq n.$

(ii) n is a **lower bound** for X iff:
$\forall M \in [M_0\rangle: |(M|X)| \geq n.$

The set X is **bounded** iff it has an upper bound.

Zero is a trivial lower bound for all X. This means that it makes no sense to ask whether X is "downwards bounded" (and thus we abbreviate "upwards bounded" to "bounded"). Usually, we are not satisfied with arbitrary bounds. Instead we want to find the best possible bounds, i.e., the minimal upper bound and the maximal lower bound.

Above we have defined upper and lower bounds for arbitrary sets of token elements. Below we consider some particular sets of token elements (e.g., those token elements that correspond to a particular place instance). This will allow us to speak about bounds for CP-nets, places and place instances. For a place instance $(p,i) \in PI$ we use $M(p,i) \in C(p)_{MS}$ to denote the marking of (p,i), i.e., the multi-set where $M(p,i)(c) = M((p,i),c)$ for all $c \in C(p)$.

Definition 4.7: Let a place instance $(p,i) \in PI$, a multi-set $m \in C(p)_{MS}$ and a non-negative integer $n \in \mathbb{N}$ be given.

(i) n is an **upper integer bound** for (p,i) iff:
$\forall M \in [M_0\rangle: |M(p,i)| \leq n$.

(ii) m is an **upper multi-set bound** for (p,i) iff:
$\forall M \in [M_0\rangle: M(p,i) \leq m$.

Lower bounds are defined analogously. The place instance (p,i) is **bounded** iff it has an upper integer bound.

We use TE(p) to denote the set of token elements that correspond to a given place p. Analogously, we use TE(p,i) to denote the set of token elements that correspond to a given place instance $(p,i) \in PI$. It is then easy to see that the integer bounds of (p,i) are identical to the bounds of TE(p,i). It can also be seen that each upper integer bound n determines an upper multi-set bound $n*TE(p,i)$, while each *finite* upper/lower multi-set bound m determines an upper/lower integer bound $|m|$.

When a bound, of one of the four types, is valid for all place instances of a given place $p \in P$, we say that the bound is valid for p. The place p is bounded iff all its place instances are bounded. Analogously, a CP-net is bounded iff all its places are bounded.

The above concepts are sufficient for many CP-nets. It is, however, rather often the case that we want to count the token elements in slightly more elaborated ways. To achieve this, we allow the modeller to define a function $F \in [\mathbb{M} \rightarrow A]$ where $(A,\leq)$ is an arbitrary set with an ordering relation (e.g., the set of integers with the ordinary less than or equal operation). We then say that an element $a \in A$ is an upper bound for the function F iff $F(M) \leq a$ for all $M \in [M_0\rangle$. Lower bounds are defined analogously.

Below we illustrate the definitions of this section by applying them to some of the CP-nets from Chaps. 1–3. These systems are rather small and simple. Thus it should be possible for the reader to verify the stated boundedness properties – without using formal proof techniques. In Chap. 5 and Vol. 2 we show how to *prove* boundedness properties in a more systematic and formal way. A similar remark applies to the examples in Sects. 4.3 – 4.5.

Data base

Let n = |DBM|.

Best upper	Multi-set	Integer
Inactive	DBM	n
Waiting	DBM	1
Performing	DBM	n−1
Unused	MES	n * (n−1)
Sent, Received, Acknowledged	MES	n−1
Passive, Active	E	1

Best lower	Multi-set	Integer
Unused	empty	$(n-1)^2$
All other places	empty	0

Notice that multi-set bounds and integer bounds supplement each other. From one of them it is often possible to deduce information which cannot be deduced from the other, and vice versa. This is, e.g., the case for the places *Waiting* and *Inactive*. For *Waiting* the upper integer bound gives us much more precise information than the upper multi-set bound. For *Inactive* it is the other way round.

Resource allocation

Let P(p) denote the set of all p-tokens, i.e., those tokens in the colour set P which have p as their first component. Analogously, P(q) denotes the q-tokens.

Best upper	Multi-set	Integer
A	3*P(q)	3
B	2*P(p) + P(q)	3
C, D, E	P(p) + P(q)	1
R	1`e	1
S	3`e	3
T	2`e	2

Best lower	Multi-set	Integer
A, B	empty	1
All other places	empty	0

Now let PR_1 be the projection function which maps a multi-set in P_{MS} into a multi-set in U_{MS} by throwing away the second component of each token colour.

For each place p∈{A,B,C,D,E} we consider the function which maps a marking M into $PR_1(M(p))$, and we then get the following bounds:

$PR_1(M(p))$	Best upper	Best lower
A	3\`q	1\`q
B	2\`p + 1\`q	1\`p
C, D, E	1\`p + 1\`q	empty

Notice that these bounds give us much more precise information than the standard multi-set bounds shown in the first table.

Telephone

Let n = |U| and assume that n is even. $UxU_{DIF} \subseteq UxU$ contains those pairs (x,y)∈UxU which satisfy x ≠ y.

Best upper	Multi-set	Integer
Inactive, Engaged, Continuous, NoTone, Short, Connected	U	n
Long, Ringing, Replaced	U	n/2
Disconnected	U	n−1
Request	UxU	n
Call, Connection	UxU_{DIF}	n/2

Best lower	Multi-set	Integer
All places	empty	0

Ring network

Let $INT_+ = \{i \in INT \mid i > 0\}$.

Best upper	Multi-set	Integer
PackNo	INT_+	1
All other places	unbounded	unbounded

Best lower	Multi-set	Integer
PackNo	empty	1
All other places	empty	0

Notice that 1 is both an upper and a lower integer bound for *PackNo*. This means that each marking has exactly one token at this place.

4.3 Home Properties

Intuitively, a home marking is a marking to which it is always possible to return. Analogously, a home space is a set of markings such that it is always possible to return to one of these:

Definition 4.8: Let a marking $M \in \mathbb{M}$ and a set of markings $X \subseteq \mathbb{M}$ be given:

(i) M is a **home marking** iff:
$\forall M' \in [M_0\rangle: M \in [M'\rangle$.

(ii) X is a **home space** iff:
$\forall M' \in [M_0\rangle: X \cap [M'\rangle \neq \emptyset$.

We use HOME to denote the set of all home markings.

It is easy to see that $M \in$ HOME iff $\{M\}$ is a home space. Notice that the existence of a home marking tells us that it is possible to reach the home marking. However, it is not guaranteed that we ever do this. In other words, there may exist infinite occurrence sequences which do not contain the home marking. A similar remark applies to home spaces.

Proposition 4.9: The following property is satisfied:
$\forall M \in \mathbb{M}: (M \in \text{HOME} \Leftrightarrow [M\rangle = \text{HOME})$.

Proof: To prove $\Leftarrow$ we notice that $M \in [M\rangle$. To prove $\Rightarrow$ we assume that M is a home marking, i.e.:

(*) $\forall M' \in [M_0\rangle: M \in [M'\rangle$,

and we then have to show that $[M\rangle = \text{HOME}$.

Case A, $[M\rangle \subseteq HOME$: Assume that $M'' \in [M\rangle$. Then we shall show that $M'' \in$ HOME, i.e.:

$\forall M' \in [M_0\rangle: M'' \in [M'\rangle$.

This follows from

$M'\ [\sigma_1\rangle\ M\ [\sigma_2\rangle\ M''$

where the existence of σ_1 is due to (*) while the existence of σ_2 is due to the assumption $M'' \in [M\rangle$.

Case B, $[M\rangle \supseteq HOME$: Assume that $M'' \in$ HOME, i.e.:

(**) $\forall M' \in [M_0\rangle: M'' \in [M'\rangle$.

Then we shall show that $M'' \in [M\rangle$. However, this is a direct consequence of (**) and the fact that $M \in [M_0\rangle$ (which follows from (*) and $M_0 \in [M_0\rangle$). □

Below we illustrate the definitions of this section by applying them to some of the CP-nets from Chaps. 1–3.

Data base

The initial marking is a home marking (and thus we have HOME = $[M_0\rangle$, which means that each reachable marking is a home marking). Moreover, we have the stronger property that each infinite occurrence sequence (starting from M_0) contains the initial marking infinitely many times.

Telephone

The initial marking is a home marking. This can be verified by proving that each sender and each recipient is able to return to *Inactive* – while removing all corresponding tokens on *Engaged, Request, Call,* and *Connection.*

Resource allocation

Let PR_1 be defined as in Sect. 4.2, and let X contain those markings which satisfy the following two properties:

$$\forall p \in \{R,S,T\}: \; M(p) = M_0(p)$$
$$\forall p \in \{A,B,C,D,E\}: \; PR_1(M(p)) = PR_1(M_0(p)).$$

Then X is a home space, and this can be verified by proving that each p-process and each q-process is able to return to its initial state (B or A). It may be a help to use some of the upper bounds found in Sect. 4.2. However, it is not easy to formulate a totally convincing proof without the use of formal methods. We shall return to this example in Chap. 5, when we discuss the use of place invariants.

Ring network

Let X contain those markings which have no tokens on *Package, Incoming* and *Outgoing*. Then X is a home space, and this can be verified by proving that all packages (created by *NewPack*) can eventually be *Received* by the designated receiver.

4.4 Liveness Properties

Intuitively, liveness tells us that a set of binding elements X remains active. This means that it is possible, for each reachable marking M', to find an occurrence sequence starting in M' and containing an element from X. It should be noted that liveness only demands that elements of X can become enabled. Thus there may be infinite occurrence sequences starting in M' and containing no elements of X.

Definition 4.10: Let a marking $M \in \mathbb{M}$ and a set of binding elements $X \subseteq BE$ be given.

(i) M is **dead** iff no binding element is enabled, i.e., iff:
$\forall x \in BE: \neg M[x\rangle$.

(ii) X is **dead** in M iff no element of X can become enabled, i.e., iff:
$\forall M' \in [M\rangle\ \forall x \in X: \neg M'[x\rangle$.

(iii) X is **live** iff there is no reachable marking in which X is dead, i.e., iff:
$\forall M' \in [M_0\rangle\ \exists M'' \in [M'\rangle\ \exists x \in X: M''[x\rangle$.

As a shorthand, we say that X is dead iff it is dead in M_0. It is easy to see that a marking M is dead iff BE is dead in M. It should be noted that live is *not* the negation of dead. Each live set of binding elements is non-dead – but the opposite is not true.

Proposition 4.11: For all X, $Y \subseteq BE$ we have:

(i) $X \supseteq Y \Rightarrow$ (X dead $\Rightarrow$ Y dead).

(ii) $X \subseteq Y \Rightarrow$ (X live $\Rightarrow$ Y live).

Proof: The proof is straightforward, and hence it is omitted. □

Above we have defined two properties, dead and live, for arbitrary sets of binding elements. We shall, however, often consider some particular sets of binding elements, and this will allow us to speak about dead and live CP-nets, transitions and transition instances. We use BE(t) to denote the set containing all those binding elements which correspond to a given transition t. We then define that t is dead iff BE(t) is dead, and that t is live iff BE(t) is live. We also define that t is **strictly live** iff the set {x} is live for all $x \in BE(t)$. Analogously, we use BE(t,i) to denote the set containing all those binding elements which correspond to a given transition instance $(t,i) \in TI$. We then define dead, live and strictly live transition instances in the same way as we defined these concepts for transitions – except that we replace BE(t) with BE(t,i). Finally, we define the three concepts for a CP-net – by considering the set of all binding elements BE. From this it follows that a CP-net is live iff it has no reachable marking which is dead.

The above concepts are sufficient for many CP-nets. However, it is rather often the case that we are not interested in a particular binding element, but in a set of closely related binding elements. As a typical example, consider the resource allocation system in Fig. 1.7. For this CP-net it does not make much sense to investigate whether a particular binding element can occur repeatedly. The problem is that each cycle of a process token increases the cycle counter, and this means that the same binding element can occur at most two or three times (depending on whether it involves a p-token or a q-token). Thus we want an easy way to consider the set of binding elements which correspond to a particular transition and to a particular binding of the variable x (while we ignore the binding of the variable i). To achieve this we allow the modeller to specify an equivalence relation $\approx$ on BE(t). We then define that a transition $t \in T$ is live with respect to $\approx$ iff each equivalence class of $\approx$ is live. Liveness with respect to an equivalence relation is also defined for transition instances and for the entire CP-net. This is done by considering equivalence relations on BE(t,i) and BE, respectively.

For those who are interested in the theoretical relationship between the different kinds of liveness, it can be noted that liveness (of a CP-net, transition or transition instance) is the same as liveness with respect to the equivalence relation where all binding elements are equivalent. Analogously, strict liveness is the same as liveness with respect to the equivalence relation where two binding elements are equivalent iff they are identical.

Above we have used the general liveness property in Def. 4.10 (iii) to define two derived properties: strict liveness and liveness with respect to an equivalence relation. In Sect. 4.5 we will define three different fairness properties: impartiality, fairness and justice. From each of these we then derive a strict property and also a property based on equivalence relations. The derived concepts are obtained from the original concepts in exactly the same way as for liveness. Analogously, we could have defined what it means that a CP-net, transition and transition instance is strictly dead and dead with respect to an equivalence relation. However, it is rather easy to see that these concepts give us nothing new – because Def. 4.10 (ii) already has a universal quantifier for x.

Below we illustrate the definitions of this section by applying them to some of the CP-nets from Chaps. 1–3.

Data base

Transitions SM and RA are strictly live, while RM and SA only are live. This can be verified by proving the following two properties:

- The initial state M_0 is a home marking.
- There exists an occurrence sequence that starts in M_0 and contains all binding elements of SM and RA and some binding elements of RM and SA.

Telephone

The CP-net is strictly live, and this can be verified in a way which is similar to that of the data base system.

Resource allocation

The CP-net is live with respect to the equivalence relation $\approx_A \subseteq BE \times BE$ where two binding elements are equivalent iff they are identical when we ignore the value bound to i. This can be verified in the following way:

- The net has the home space X found in Sect. 4.3.
- For each marking $M \in X$ there exists an occurrence sequence starting in M and containing a binding element from each equivalence class of $\approx_A$.

Moreover, it can be seen that a binding element (t,<x=v,i=n>) is dead in a marking $M \in [M_0\rangle$ iff M does not contain a token element:

- (p,(v,m)) where that m < n, or
- (p,(v,n)) where p is positioned above t (in the layout of Fig. 1.7).

Ring network

The CP-net is live with respect to the equivalence relation $\approx_B \subseteq BE \times BE$ where two binding elements are equivalent iff they are identical when we ignore the value bound to n and the no-component of the value bound to p. This can be verified in a way which is similar to the one used for the resource allocation system.

4.5 Fairness Properties

Intuitively, fairness tells us how often the different binding elements occur. In this section we only consider occurrence sequences that start in a reachable marking. Let $X \subseteq BE$ be a set of binding elements and $\sigma \in OSI$ an infinite occurrence sequence of the form:

$$\sigma = M_1 [Y_1\rangle M_2 [Y_2\rangle M_3 \ldots$$

For each $i \in \mathbb{N}_+$, we use $EN_{X,i}(\sigma)$ to denote the number of elements from X which are enabled in the marking M_i (when an element is concurrently enabled with itself this is reflected in the count). Analogously, we use $OC_{X,i}(\sigma)$ to denote the number of elements from X which occur in the step Y_i (when an element occurs concurrently with itself this is reflected in the count).

We use $EN_X(\sigma)$ and $OC_X(\sigma)$ to denote the total number of enablings/occurrences in σ, i.e.:

$$EN_X(\sigma) = \sum_{i \in \mathbb{N}_+} EN_{X,i}(\sigma).$$

$$OC_X(\sigma) = \sum_{i \in \mathbb{N}_+} OC_{X,i}(\sigma).$$

Since all elements in the two sums are non-negative integers, it is easy to see that the sums must be convergent – either to an element of $\mathbb{N}$ or to ∞. Now we define three different fairness properties:

Definition 4.12: Let $X \subseteq BE$ be a set of binding elements and $\sigma \in OSI$ an infinite occurrence sequence.

(i) X is **impartial** for σ iff it has infinitely many occurrences, i.e., iff:
$OC_X(\sigma) = \infty$.

(ii) X is **fair** for σ iff an infinite number of enablings implies an infinite number of occurrences, i.e., iff:
$EN_X(\sigma) = \infty \Rightarrow OC_X(\sigma) = \infty$.

(iii) X is **just** for σ iff a persistent enabling implies an occurrence, i.e., iff:
$\forall i \in \mathbb{N}_+: [EN_{X,i}(\sigma) \neq 0 \Rightarrow \exists k \geq i: [EN_{X,k}(\sigma) = 0 \vee OC_{X,k}(\sigma) \neq 0]]$.

When X is impartial for all infinite occurrence sequences of the given CP-net, we say that X is impartial. Analogous definitions are made for fair and just.

Proposition 4.13: For all $X, Y, X_1, X_2, \ldots, X_n \subseteq BE$ and all infinite occurrence sequences $\sigma \in OSI$ we have:

(i) X is impartial for $\sigma \Rightarrow$ X is fair for $\sigma \Rightarrow$ X is just for σ.
(ii) $X \subseteq Y \Rightarrow$ (X is impartial for $\sigma \Rightarrow$ Y is impartial for σ).
(iii) $X_1 \cup X_2 \cup \ldots \cup X_n = X \Rightarrow$
$((\forall i \in 1..n$: X_i is fair for $\sigma) \Rightarrow$ X is fair for σ).

The same three properties are satisfied when we omit "for σ".

Proof: The proof of the left implication in (i) and the proof of (ii) are straightforward, and hence they are omitted.

To prove the right implication in (i) assume that X is fair, and also assume that $EN_{X,i}(\sigma) \neq 0$. If we can find $k \geq i$ such that $EN_{X,k}(\sigma) = 0$ we have finished the proof. Otherwise $EN_{X,k}(\sigma) \neq 0$ for all $k \geq i$ and thus we have $EN_X(\sigma) = \infty$. Then the fairness of X implies that $OC_X(\sigma) = \infty$, and thus we can find $k \geq i$ such that $OC_{X,k}(\sigma) \neq 0$.

To prove (iii) assume that $X_1, X_2, \ldots, X_n$ are fair, and also assume that $EN_X(\sigma) = \infty$. From $X_1 \cup X_2 \cup \ldots \cup X_n = X$ it follows that there exists an $i \in 1..n$ such that $EN_{X_i}(\sigma) = \infty$. Then the fairness of X_i implies that $OC_{X_i}(\sigma) = \infty$, and finally we use $X_i \subseteq X$ to conclude that $OC_X(\sigma) = \infty$. □

It should be noted that neither (ii) nor (iii) are correct when impartial and fair are replaced by just. Now we define what it means that a CP-net, transition and transition instance is impartial, fair and just (for a given occurrence sequence, and in general). This is done by considering BE, BE(t) and BE(t,i), respectively. We also define **strict impartiality, strict fairness** and **strict justice**. This is done by considering all sets {x} where x belongs to BE, BE(t) and BE(t,i), respectively. Finally, we define impartiality, fairness and justice with respect to an equivalence relation. This is done by considering all the equivalence classes in the given relation.

The above concepts are sufficient for many CP-nets. It is, however, rather often the case that a CP-net has some particular binding elements which may be repeated indefinitely. As a typical example, consider the ring network in Figs. 3.1 and 3.2. For this CP-net it does not make much sense to investigate whether *Send* is impartial, fair or just. The problem is that there exist infinite occurrence sequences which only contain elements of BE(*NewPack*). To solve this problem we allow the modeller to remove certain binding elements from the considered occurrence sequences – before the properties of Def. 4.12 are checked. This means that each occurrence sequence σ is replaced by a **pruned occurrence sequence** $\sigma \backslash Y$ in which all elements of Y are removed from the steps – while the markings are unaltered. Notice that some of the pruned steps in $\sigma \backslash Y$ may become empty, because all elements of the original step belonged to Y.

A pruned occurrence sequence $\sigma \backslash Y$ is **infinite** iff it contains an infinite number of binding elements. A set $X \subseteq BE$ is impartial, fair or just, for a pruned occurrence sequence $\sigma\backslash Y$, iff the properties of Def. 4.12 are satisfied – when we replace $EN_{X,i}(\sigma)$, $OC_{X,i}(\sigma)$, $EN_X(\sigma)$ and $OC_X(\sigma)$ by $EN_{X,i}(\sigma\backslash Y)$, $OC_{X,i}(\sigma\backslash Y)$, $EN_X(\sigma\backslash Y)$ and $OC_X(\sigma\backslash Y)$, respectively.

Definition 4.14: Let $X,Y \subseteq BE$ be two sets of binding elements and $\sigma \in OSI$ an infinite occurrence sequence. We then say that X is **impartial** for σ with Y **ignored** iff:

$$\sigma\backslash Y \text{ infinite} \Rightarrow X \text{ is impartial for } \sigma\backslash Y.$$

When X with Y ignored is impartial for all occurrence sequences of the given CP-net, we say that X is impartial with Y ignored.

Analogous definitions are made for fair and just.

Now we repeat all the definitions of the derived concepts (i.e., the strict properties and the properties based on equivalence relations). As an example, we can then say that t_1 is strictly fair for σ with t_2 and t_3 ignored, or we can say that a CP-net is fair with respect to $\approx$ when a set of transition instances is ignored. It should be noted that Prop. 4.13 remains true when we have an ignored set of binding elements. The proof of this is straightforward, and thus it is omitted. Below we illustrate the definitions of this section by applying them to some of the CP-nets from Chaps. 1–3.

Data base

Each transition is impartial. *Update and Send Messages* are strictly just, while the other three transitions are strictly fair.

Telephone

The set of bindings elements which correspond to the "replace" transitions is impartial. *LiftInac@Dialing#3* is impartial when we ignore *LiftRepl@BreakRec#6* and *RepConRec@BreakRec#6*.

Resource allocation

T2, T3, T4 and T5 are impartial. Next let us consider equivalence classes of $\approx_A$ (cf. Sect. 4.4). Each of the six classes which correspond to T3, T4 and T5 is fair. Each of the four classes where x is bound to p is impartial, when the elements of the other five classes are ignored (and a similar statement is true for the five classes where x is bound to q).

Ring network

NewPack is impartial. *Send* is impartial when *NewPack* is ignored.

Bibliographical Remarks

Most of the properties defined in Sects. 4.1–4.4 are straightforward generalizations of properties defined for low-level Petri nets (and for many other kinds of Petri nets). They can be found in most introductory texts on Petri nets, e.g., many of those mentioned in the bibliographical remarks of Chap. 1. It should, however, be noted that many Petri net papers say that a net is live iff each individual transition is live, while we say that a net is live iff BE is live. The latter is a much weaker property.

The fairness properties of Sect. 4.5 are inspired by the work of P. Chrzastowski-Wachtel in [24]. His definitions are given for PT-nets, and they build upon [66]. A definition of the net topology mentioned in Sect. 4.1 can be found in [38].

Exercises

Exercise 4.1.
This exercise deals with the static properties defined in Sect. 4.1.

(a) Prove the static properties which we have postulated for the CP-nets in Chaps. 1–3. See the end of Sect. 4.1.

(b) Find an open subnet which cannot be induced by a set of transitions, and find a closed subnet which cannot be induced by a set of places.

(c) Consider the static properties of the CP-nets for the producer/consumer system in Exercise 1.3.

(d) Consider the static properties of the CP-nets for the philosopher system in Exercise 1.6.

(e) Consider the static properties of the CP-nets for the master/slave system in Exercise 1.9.

Exercise 4.2.
This exercise deals with the boundedness properties defined in Sect. 4.2.

(a) Prove the bounds which we have postulated for the CP-nets in Chaps. 1–3. See the end of Sect. 4.2.

(b) What happens to the bounds of the telephone system when U has an odd number of elements?

(c) Consider the boundedness properties of the CP-nets for the producer/consumer system in Exercise 1.3.

(d) Consider the boundedness properties of the CP-nets for the philosopher system in Exercise 1.6.

(e) Consider the boundedness properties of the CP-nets for the master/slave system in Exercise 1.9.

Exercise 4.3.
This exercise deals with the home properties defined in Sect. 4.3.

(a) Prove the home properties which we have postulated for the CP-nets in Chaps. 1–3. See the end of Sect. 4.3.

(b) Consider the home properties of the CP-nets for the producer/consumer system in Exercise 1.3.

(c) Consider the home properties of the CP-nets for the philosopher system in Exercise 1.6.

(d) Consider the home properties of the CP-nets for the master/slave system in Exercise 1.9.

Exercise 4.4.
This exercise deals with the liveness properties defined in Sect. 4.4.

(a) Prove the liveness properties which we have postulated for the CP-nets in Chaps. 1–3. See the end of Sect. 4.4.

(b) Find a set of binding elements which is non-dead without being live.

(c) Consider the liveness properties of the CP-nets for the producer/consumer system in Exercise 1.3.

(d) Consider the liveness properties of the CP-nets for the philosopher system in Exercise 1.6.

(e) Consider the liveness properties of the CP-nets for the master/slave system in Exercise 1.9.

Exercise 4.5.
This exercise deals with the fairness properties defined in Sect. 4.5.

(a) Prove the fairness properties which we have postulated for the CP-nets in Chaps. 1–3. See the end of Sect. 4.5.

(b) Find sets of binding elements X and Y which demonstrate that Prop. 4.13 (ii) is false when impartial is replaced by fair or by just.

(c) Find sets of binding elements which demonstrate that Prop. 4.13 (iii) is false when fair is replaced by just.

(d) Consider the fairness properties of the CP-nets for the producer/consumer system in Exercise 1.3.

(e) Consider the fairness properties of the CP-nets for the philosopher system in Exercise 1.6.

(f) Consider the fairness properties of the CP-nets for the master/slave system in Exercise 1.9.

Exercise 4.6.
Consider the following CP-net which models a small process control system. It contains a set of processes, PROC, which *Measure* the state of a subsystem and *Report* the *Results* to a central *Controller*, which may raise *Alarms*. Finally, there is a process, which can sometimes *Fix* the problems.

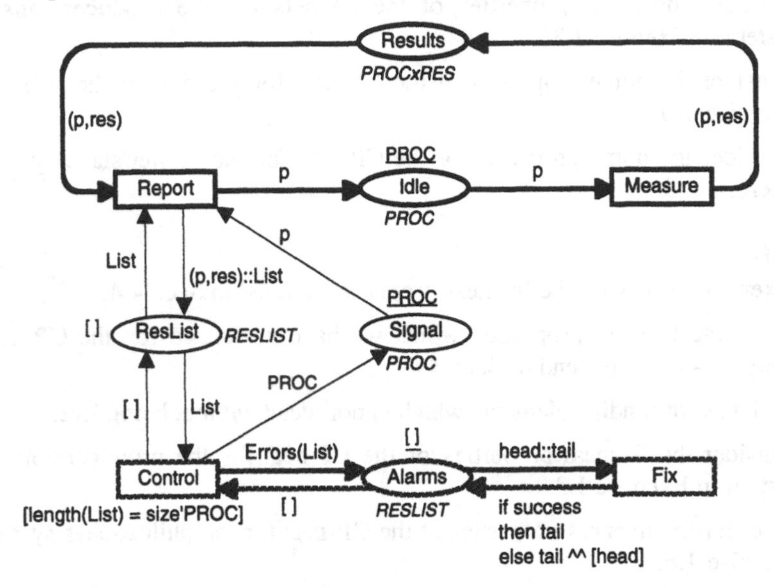

There is a set of possible results, RES, of which some cause alarms while others are okay. The function *Errors* returns a list containing those elements (of its argument) which cause an *Alarm*. The guard in the lower left corner tests whether all processes have reported their results. This is done by comparing the length of the result list and the size of PROC. The variable *success* is of type boolean. When it is bound to true it is possible to fix the error; otherwise the corresponding alarm is moved to the end of the alarm list (and may be fixed at some later point).

(a) Finish the CP-net by writing all the necessary declarations. Make a simulation of the constructed CP-net.

(b) Investigate the boundedness properties of the process control system.

(c) Investigate the home properties of the process control system.

(d) Investigate the liveness properties of the process control system.

(e) Investigate the fairness properties of the process control system.

Chapter 5

Formal Analysis of Coloured Petri Nets

This chapter contains an informal introduction to the analysis of CP-nets. The most straightforward kind of analysis is simulation, which in many respects is similar to the testing and execution of a program. A good CPN simulator is analogous to a good program debugger. It assists the user in a careful examination of some of the possible execution sequences.

Simulation is extremely useful for the understanding and debugging of a CP-net, in particular during the design and early validation of a large system. However, it is obvious that, by means of simulation, it is impossible to obtain a complete proof of dynamic properties of CP-nets (unless the nets or the properties are trivial). Thus it is very important that there also exist a number of more formal analysis methods (i.e., methods which are based on mathematical proof techniques).

In the present chapter we introduce the basic ideas behind the most important of the formal analysis methods. Our description is intuitive, imprecise and without proofs. However, most of the content of Vol. 2 is devoted to the formal analysis methods – there we give the precise unambiguous definitions, prove the correctness of the various techniques, and explain how to use the methods in practice. In the present chapter we hope to motivate our readers to study also the more mathematical and complex aspects of CP-nets. The formal analysis methods are, in our opinion, an indispensable complement to the more straightforward and intuitive simulation possibilities.

Section 5.1 introduces occurrence graphs, while Sect. 5.2 deals with place and transition invariants and Sect. 5.3 with reduction rules. Finally, Sect. 5.4 discusses performance analysis (i.e., analysis of the speed by which a system operates). In Chap. 6 we describe computer tools that support the use of CP-nets, and we then also consider tools supporting the formal analysis methods.

5.1 Occurrence Graphs

The basic idea behind occurrence graphs is to construct a graph containing a node for each reachable marking and an arc for each occurring binding element. Obviously such a graph may become very large, even for small CP-nets. As an example, let us consider the resource allocation system in Fig. 1.7. Due to the cycle counters this net has an infinite number of reachable markings and thus an infinite occurrence graph. We can, however, simplify the CP-net by omitting the cycle counters. Then we use U as colour set for all places A–E, and in the net inscriptions we replace (x,i) by x, (p,i+1) by p, (q,i+1) by q, (p,0) by p and (q,0) by q. It is easy to check that the cycle counters form an isolated part of the original CP-net – in the sense that they neither influence the enabling nor the effect of an occurrence (except that they determine the values of new cycle counters). This means that the simplified net has a similar behaviour to the original net. For each occurrence sequence in one of the CP-nets there is a corresponding occurrence sequence in the other. Hence we can get information about the dynamic properties of the original net by constructing an occurrence graph for the simplified net.

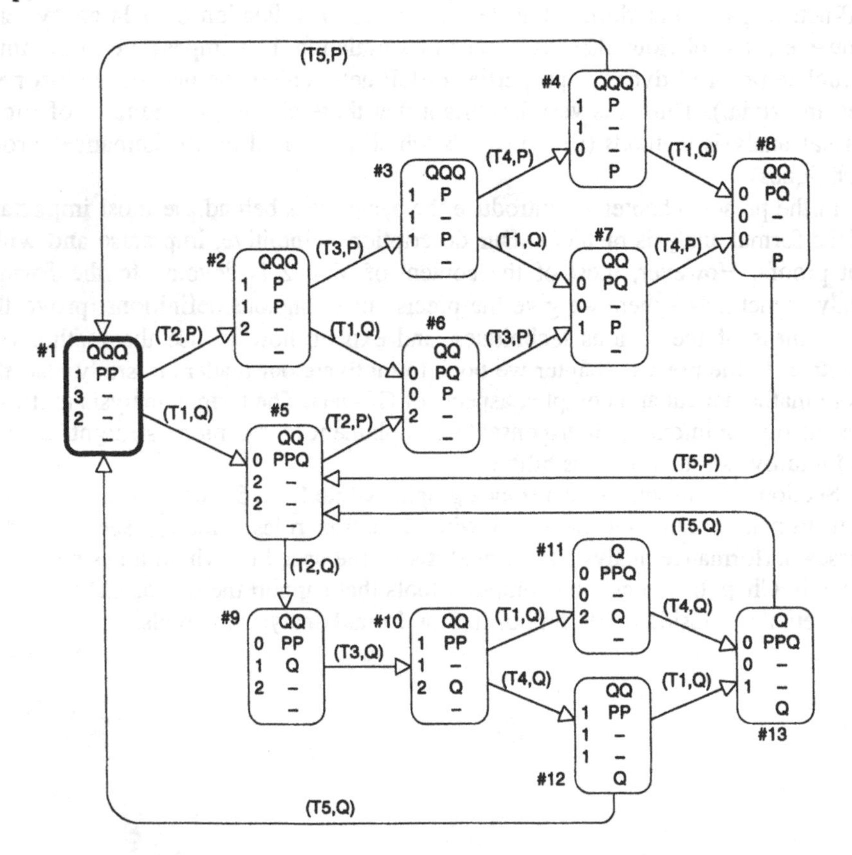

Fig. 5.1. Full occurrence graph for a simplified version of the resource allocation system

Such a graph is shown in Fig. 5.1 – it is called a **full occurrence graph**. Each node represents a marking, and the content of this marking is described by the text inscription of the node. The left column contains the number of e-tokens on the three resource places R–T, while the right column contains the p-tokens and q-tokens on the places A–E (to improve readability we use P and Q, instead of p and q). The node with thick border line represents the initial marking. Each arc represents the occurrence of a binding element, and the content of this binding element is described by the text attached to the arc. The left component of the pair indicates the transition, while the right one indicates the value bound to x. From Fig. 5.1 it is straightforward to verify all the dynamic properties postulated for the resource allocation system in Chap. 4.

Notice that in occurrence graphs we usually omit arcs that correspond to steps containing more than one binding element. Otherwise, we would, e.g., have had an arc from node #1 to node #6, with the inscription 1`(T2,P)+1`(T1,Q). Such arcs would give us information about the concurrency between binding elements, but they are not necessary for the verification of the dynamic properties defined in Chap. 4.

Even for a small occurrence graph, like the one in Fig. 5.1, the construction and investigation are tedious and error-prone. In practice it is not unusual to handle CP-nets which have occurrence graphs containing more than 100,000 nodes (and many CP-nets have millions of markings). Thus it is obvious that we need to be able to construct and investigate the occurrence graphs by means of a computer, and also that we want to develop techniques by which we can construct reduced occurrence graphs without losing too much information. As shown below, there are several different ways to obtain such a reduction. For all of them, it is important to notice that the reduced occurrence graph can be obtained directly, i.e., calculated without first constructing the full occurrence graph.

Symmetrical markings

This reduction method exploits the symmetries which we often have in systems modelled by CP-nets. As an example, let us consider the data base system in Fig. 1.13. Let M_1 be the reachable marking in which manager d_1 is *Waiting* (while the remaining managers are *Performing*), and let M_2 be the reachable marking in which manager d_2 is *Waiting* (while the remaining managers are *Performing*). Intuitively, there is very little difference between the two markings. They are symmetrical, in the sense that one of them can be obtained from the other by a consistent renaming of the managers (interchanging d_1 with d_2 in all token colours). It can be verified that each transition in the data base system handles all colours in a similar, i.e., symmetrical way (this is a static property which can be checked by a local inspection of each individual transition). Hence it follows that each occurrence sequence starting from M_1 has a symmetrical occurrence sequence starting from M_2, and vice versa. The two sequences are obtained from each other by interchanging d_1 with d_2 (in all markings and all steps).

The observations above mean that it makes sense to construct a reduced occurrence graph which only contains a node for each class of symmetrical markings. This means that M_1 and M_2 (together with two other markings M_3 and M_4) are represented by a single node. Let us assume that for this node we choose M_1 as the representative marking (attached to the node). Then we notice that the colours d_2, d_3 and d_4 appear in M_1 in a symmetrical way. This means that any interchange of these three colours maps M_1 into itself. Next consider the three binding elements (SA,<s=d_1,r=d_2>), (SA,<s=d_1,r=d_3>) and (SA,<s=d_1,r=d_4>) which all are enabled in M_1. These binding elements are symmetrical, in the sense that they can be obtained from each other by interchanging d_2, d_3 and d_4. From the symmetry of M_1, the symmetry of the three binding elements and the symmetry of each transition, it can be proved that the three binding elements, when they occur in M_1, will yield three markings which are known to be symmetrical. Thus it is sufficient to consider one of the binding elements.

The reduced graph for the data base system is shown in Fig. 5.2. Each node represents a class of markings, and the content of one of these is described by the text inscription of the node. The first row indicates the markings of the places *Waiting, Inactive, Performing*, *Active* and *Passive*, while the next three lines indicate the marking of *Sent, Received* and *Acknowledged*. For brevity we write i instead of d_i and we omit the marking of *Unused*. Each arc represents a class of occurring binding elements, and the content of one of these is described by the text attached to the arc. For brevity we write (SM,i) and (RM,i,k) instead of (SM,<s=d_i>) and (RM,<s=d_i,r=d_k>), and analogously for SA and RA. Node #8 represents the markings M_1–M_4 and its outgoing arc represents the three binding elements considered above.

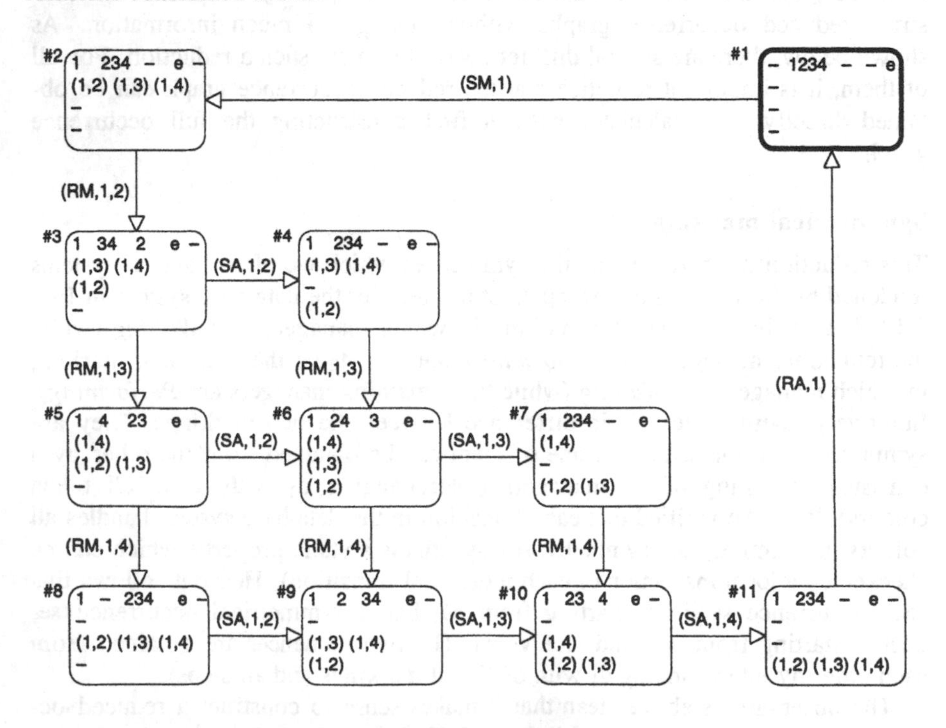

Fig. 5.2. Reduced occurrence graph for the data base system

In Vol. 2 we shall give a formal definition of occurrence graphs with symmetrical markings. This will involve a number of well-known mathematical concepts, e.g., algebraic groups and equivalence classes. We shall then prove that the reduced graph contains exactly the same information as the full occurrence graph. This means that the reduced graph can be used to investigate all those system properties which can be investigated by means of the full graph. The reduced graph is a folded version of the full occurrence graph, in a similar way as a CP-net is a folding of the equivalent PT-net. The full occurrence graph can be constructed from the reduced graph, but this is never necessary. Instead we perform the analysis directly on the reduced graph, and this turns out to be much more efficient (compared to the analysis of the full occurrence graph).

In Vol. 2, we will also prove that the reduced occurrence graph for the data base system (for all $n \geq 3$) contains $1+n*(n+1)/2$ nodes and $2+(n-1)*n$ arcs, while the full occurrence graph contains $1+n*3^{n-1}$ nodes and $2n+2*(n-1)*n*3^{n-2}$ arcs. This means that the reduced graph grows with quadratic speed, while the full graph grows exponentially. For $n = 4$ the reduced graph has 11 nodes and 14 arcs, while the full graph has 109 nodes and 224 arcs. For $n = 10$ the reduced graph has 56 nodes and 92 arcs, while the full graph has 196,831 nodes and 1,181,000 arcs.

Stubborn sets

A second reduction possibility builds upon the observation that a CP-net often has a number of occurrence sequences where the steps are identical, except for the order in which they occur. Let a marking M have n concurrently enabled binding elements (which are different). Then we can sort these elements in n! different ways. Each of these determines a possible occurrence sequence starting in M. For $n = 4$ we have 24 different possibilities. For $n = 10$ we have 3,628,800 possibilities. However, the total effects of all these occurrence sequences are the same, in the sense that they all lead to the same marking. Thus it is natural to ask whether it is really necessary to develop all of them, and fortunately it turns out that this is not the case. Instead, for each reachable marking, we calculate a so-called stubborn set, and we use this set to tell us which of the enabled binding elements we need to investigate, i.e., to let occur. The remaining binding elements are, by the definition of stubborn sets, guaranteed to remain enabled, and thus they can occur in the next marking (or in a later one). The use of stubborn sets often gives a very significant reduction of the number of nodes and the number of arcs – in particular when the modelled system contains a large number of relatively independent processes. Unfortunately, with the stubborn set method it is sometimes necessary to construct several different occurrence graphs (because the definition of stubborn sets depends upon those properties which we want to investigate).

The use of stubborn sets can be combined with the use of symmetrical markings. The two methods are orthogonal, in the sense that symmetrical markings are useful when we have a number of symmetrical processes, while stubborn sets are useful when we have a number of concurrent processes. When the processes are both symmetrical and concurrent we can use both methods, simultaneously.

This often gives us an improved reduction. As an example, the data base system gets a reduced occurrence graph, which is identical to Fig. 5.2 except that the interior nodes, #4, #6 and #7, disappear. At a first glance this may not seem to be a significant reduction. However, it implies that the graph now only contains 2*n nodes and 2*n arcs. Hence it grows with linear speed (instead of quadratic). For n = 10 we save 64% of the nodes and 78% of the arcs (compared to situation where we only use symmetrical markings).

Covering markings

A third reduction possibility is to look for occurrence sequences which lead from a reachable marking M_1 to a covering marking M_2 (i.e., a marking which is strictly larger than M_1). The total effect of such a sequence is to add tokens, and thus it is easy to see that the steps can be repeated as many times as we want. This means that some of the token elements can get an arbitrarily high coefficient, because each repetition of the sequence increases the coefficient. Such coefficients are replaced by ∞ and we then have an occurrence graph where some nodes represent many different markings (which all are identical except for those token elements which have a ∞ coefficient). For a CP-net with finite colour sets it can be proved that reduction by means of covering markings always yields a finite occurrence graph. However, the method has some drawbacks. First of all it only gives a reduction for unbounded systems (and most practical systems are bounded). Secondly, so much information is lost by the reduction that several important properties (e.g., liveness and reachability) are no longer fully decidable.

It is possible to use covering markings together with symmetrical markings, but this implies that the theoretical justification of the method becomes much more complex (compared to the situation where only symmetrical markings are used).

Proof rules

All the above methods have originally been proposed for PT-nets or for non-hierarchical CP-nets. However, it turns out to be straightforward to extend them to hierarchical CP-nets – due to the behavioural equivalences described in Sects. 2.4 and 3.5.

When an occurrence graph has been constructed it can be used to prove properties about the modelled system. This is done by applying a set of proof rules, i.e., theorems which allow the modeller to deduce properties of the given CP-net from properties of the occurrence graph. The construction and analysis of occurrence graphs can be totally automated. This means that a modeller can use the method and interpret the results without having much knowledge about the underlying mathematics. For bounded systems a large number of questions can be fully answered. Deadlocks, reachability and marking bounds can be decided by a simple search through the nodes of the occurrence graph, while liveness and home markings can be decided by constructing and inspecting the strongly connected components (to which we shall return in Sect. 6.3). This means that occurrence graphs constitute a straightforward way to debug new CP-nets – because trivial errors such as the omission of an arc or a wrong arc expression often mean that some of the expected system properties are dramatically changed.

One problem with occurrence graphs is the fact that it is necessary to fix all system parameters (e.g., the number of managers in the data base system) before an occurrence graph can be constructed. This means that we always find properties which are specific to the chosen values of the system parameters. In practice the problem isn't that big. When we understand how a data base system behaves for a few managers, we also know a lot about how it behaves when there are more. This is of course only true when we talk about the logical correctness of a system, and not when we speak about the performance. A more important problem is the fact that the occurrence graphs often become so large that they cannot be constructed – even when the above reduction methods are applied. Hence we often have to simplify a CP-net or analyse the different parts of it separately.

5.2 Place and Transition Invariants

The basic idea behind place invariants is to construct equations which are satisfied for all reachable markings. As an example, let us consider the resource allocation system in Fig. 1.7. From our intuition about this system we expect that the places A–E always have five tokens – two of which represent p-processes, while the remaining three represent q-processes. This is expressed by the following equation, which is satisfied for all reachable markings M. PR_1 is the projection function defined in Sect. 4.2.

$$PR_1(M(A)+M(B)+M(C)+M(D)+M(E)) = 2`p+3`q.$$

It is also straightforward to see that the use of the r-resource guarantees that we cannot have two q-processes in states B or C, simultaneously. This can be expressed as follows:

$$M(R) + Q_C(M(B) + M(C)) = 1`e.$$

The function Q_C *counts* the number of q-tokens. It maps each multi-set $m \in P_{MS}$ into the E multi-set which has as many e-tokens as m has q-tokens. P_C is defined analogously. As examples, $P_C(2`(p,2)+1`(q,1)) = 2`e$ and $Q_C(1`(q,1)+3`(q,5)) = 4`e$. Now let us be a little bit more general. We notice that each of the above equations can be written in the following form:

$$(*) \quad \sum_{p \in P} W_p(M(p)) = \sum_{p \in P} W_p(M_0(p)) \quad \in A_{MS}$$

where $A \in \Sigma$ is an arbitrary colour set and $\{W_p\}_{p \in P}$ a set of **weights** such that $W_p \in [C(p)_{MS} \rightarrow A_{MS}]$ for all $p \in P$. It should be noticed that many of the weights are identity functions or zero functions (mapping each multi-set into the empty multi-set). Now let us consider the weights more carefully. We then notice that each weight $W_p \in [C(p)_{MS} \rightarrow A_{MS}]$ is defined in such a way that:

$$W_p(m_1+m_2) = W_p(m_1)+W_p(m_2)$$

for all multi-sets $m_1, m_2 \in C(p)_{MS}$. A function with this property is said to be **linear**, and it can be proved that such a function is uniquely determined from its values on those multi-sets which have a single element.

From the linearity of the weights and from the enabling rule of CP-nets it can be proved that the following property is sufficient to guarantee that (*) is satisfied for all reachable markings:

$$(**) \quad \forall(t,b) \in BE: \sum_{p \in In(t)} W_p(E(p,t)\langle b \rangle) = \sum_{p \in Out(t)} W_p(E(t,p)\langle b \rangle).$$

The intuition behind (**) is to check that each binding element removes – when the weights are taken into account – a set of tokens that is identical to the set of tokens which is added. The above property is also a necessary condition for (*) unless there exist dead binding elements.

It should be noticed that (**) is a static property which can be checked without considering the set of reachable markings. It is also a local property which can be checked separately for each transition. These two properties makes (**) much easier to verify than (*).

In the discussion above we have used linear functions mapping multi-sets into multi-sets, and we have assumed that the summations in (*) and (**) denote multi-set addition. For the resource allocation system this is sufficient, but in general it turns out to be very useful also to allow some of the weights to be subtracted from each other (i.e., to allow a place to have a "negative" weight). The easiest and most general way to obtain this is to replace multi-sets by weighted sets. A **weighted set** (over a non-empty set S) is defined in exactly the same way as we have defined a multi-set – except that we now also allow negative coefficients. This means that two weighted sets (over the same set S) can always be subtracted from each other, and it also means that scalar-multiplication with negative integers can be allowed. The set of all weighted sets over S is denoted by S_{WS}. Weighted sets have properties which are analogous to Prop. 2.3, and we can define linear functions mapping weighted sets into weighted sets – in exactly the same way we did it for multi-sets. A formal definition of weighted sets will be given in Vol. 2.

Now let us illustrate how invariants can be used to prove dynamic properties of the corresponding CP-net. To do this we again consider the resource allocation system where we have the invariants shown below (and many others). The function $PQ_C = P_C + Q_C$ maps each multi-set $m \in P_{MS}$ into the E multi-set which has the same number of tokens as m. This means that we have, e.g., $PQ_C(2`(p,2)+1`(q,1)) = 3`e$ and $PQ_C(1`(q,1)+3`(q,5)) = 4`e$. For brevity, and improved readability, we omit M(), and this means that we write $W_p(p)$ instead of $W_p(M(p))$. PI_S and PI_T describe the use of s-resources and the use of t-resources, respectively. They are analogous to PI_R.

PI_P	$PR_1(A+B+C+D+E) = 2`p+3`q$
PI_R	$R+Q_C(B+C) = 1`e$
PI_S	$S+Q_C(B)+2*PQ_C(C+D+E) = 3`e$
PI_T	$T+P_C(D)+(PQ_C+P_C)(E) = 2`e$

Let us first show that the invariants can be used to prove the upper bounds postulated in Sect. 4.2. The bounds for R, S, and T follow from PI_R, PI_S and PI_T, respectively. The bounds for A follow from PI_P (and the observation that A can never contain p-tokens). The bounds for B follow from PI_P (which tell us that there can be at most two p-tokens, and from PI_R (which tell us that there can be at most one q-token). The bounds for C, D, and E follow from PI_S (which also tells us that at most one of these places has a token). It is straightforward to construct occurrence sequences which show that the bounds are minimal, and hence we omit this part of the proof. The invariants can also be used to verify the lower bounds (cf. Exercise 5.3).

Next let us show that the CP-net is live. *The proof is by contradiction:* Let us assume that we have a reachable marking M which is dead, i.e., has no enabled transitions. From PI_P we know that M has two p-tokens and three q-tokens, distributed on the places A–E. Now let us investigate in more detail where these tokens can be positioned.

Case a: Assume that there are tokens on E: Then T5 is enabled and we have a contradiction (with our assumption of no enabled transition).

Case b: Assume that there are tokens on C and/or D (and no tokens on E): From PI_S it follows that there can be at most one such token and then PI_T tells that there is at least one e-token on T (because $P_C(D) \leq 1`e$ and $(PQ_C+P_C)(E) = \emptyset$). Thus T3 or T4 can occur.

Case c: Assume that there are tokens on B (and no tokens on C, D and E): From PI_R it follows that there can be at most one q-token on B and then PI_S tells us that there is at least two e-tokens on S (because $Q_C(B) \leq 1`e$ and $2*PQ_C(C+D+E) = \emptyset$). Thus T2 can occur.

Now we have shown that it is impossible to position the two p-tokens on the places B–E (without violating our assumption of no enabled transition). It is also obvious that the p-tokens cannot be on the place A. Thus we conclude that our assumption must be invalid, and hence the CP-net is live. From the liveness and the cyclic structure of the CP-net, it is possible to prove several other dynamic properties, e.g., that the CP-net has the home space given in Sect. 4.3 and is live with respect to the equivalence relation given in Sect. 4.4.

In Vol. 2 we shall show how to find invariants, i.e., equations which are on the form (*) and are satisfied in all reachable markings. This will be done by means of (**) and it will involve a number of well-known mathematical concepts, e.g., lambda calculus and homogeneous matrix equations. It should, however, be noticed that we seldom have to start the calculation of invariants from scratch. Usually we have, a priori, some sets of weights, which we expect to determine invariants. Such potential invariants may be derived, e.g., from the system specification and from the modeller's knowledge of the expected system properties. The invariants may be specified during the analysis of the CP-net. It is, however, much more useful (and also easier) to specify the invariants while the CP-net is being created. This means that we construct the invariants as an integrated part of the design (in a similar way as a good programmer specifies a loop invariant at the moment he creates the loop). For this kind of use, it is important to notice that the check of invariants is constructive. From the binding elements that violate (**) we know where in the CP-net the problems are. This means that it is often rather straightforward to figure out how to modify the CP-net or the weights so that we obtain valid invariants.

Transition invariants are the duals of place invariants, and this means that we attach a weight to each transition. Intuitively, a transition invariant characterizes a set of occurrence sequences that have no total effect, i.e., have the same start and end marking. Transition invariants can be calculated in a way which is similar to that of place invariants – but it is also in this case easier and more useful to construct the invariants during the creation of the CP-net. Transition invariants are used for similar purposes as place invariants, i.e., to investigate the dynamic properties of CP-nets. It is straightforward to extend both kinds of invariants to hierarchical CP-nets. We then attach a weight to each place instance group (for place invariants) and a weight to each transition instance (for transition invariants).

Analysis by means of place and transition invariants has several attractive properties. First of all, it is possible to obtain an invariant for a hierarchical CP-net by composing invariants of the individual pages. This means that it is much easier to use invariants for large systems – without encountering the same kind of complexity problems as we have for occurrence graphs. Secondly, we can find invariants without fixing the system parameters, and hence we can obtain general properties which are independent of the system parameters. Thirdly, we can construct the invariants during the design of a system and this will usually lead to an improved design. The main drawback of invariants analysis is the fact that it requires skills which are considerably higher (and more mathematical) than those required by the other analysis methods. This means that it is more difficult to use invariants in industrial system development.

5.3 Reduction Rules

CP-nets can also be analysed by means of reduction, where the basic idea is as follows. First we choose one or more types of properties which we want to investigate (e.g., liveness and boundedness). Then we define a set of reduction rules by which we can simplify CP-nets – without changing those properties which we are investigating. Usually, the rules are local, in the sense that each of them allow a subnet to be replaced by another subnet.

A typical rule is sketched in Fig. 5.3. The reduction rule specifies the two subnets. In our example, we replace a subnet which has a place Q, two transitions T1 and T2, and two *identical* arc expressions Expr, by a subnet with only one transition T3. It is also specified how the new subnet is inserted in the original net. In our example, T3 gets an input arc for each input arc of T1 and an output arc for each output arc of T2. Each new arc has the same arc expression and the same place as the corresponding removed arc. The guard of T3 is identical to that of T1. Finally, it may be specified that the rule can only be used when certain prerequisites are fulfilled. In our example, we require that Q, T1 and T2 only have those arcs that are shown. Q must have an empty initial marking, and the guard of T2 must be true. We also require that the three nodes in the removed subnet are ordinary nodes (i.e., that they are not port, socket, fusion or substitution nodes). Finally, it is necessary to require that the arc expression, Expr, fulfils certain properties (cf. Exercise 5.5).

When we have defined a set of reduction rules, we must show that the set is sound. This means that we prove that the rules never change the set of properties which we are investigating. Finally, we can apply the reduction rules to a CP-net which we want to analyse. This gives us a reduced CP-net which is known to have the same properties as the original net. Hence we can investigate the reduced net instead of the original CP-net. A good set of reduction rules must (in addition to being sound) also be powerful. It must yield a significant reduction, i.e., produce a net which it is much easier to investigate than the original net. Many of the known reduction techniques for CP-nets are direct translations of reductions rules developed for PT-nets. The reduced net is sometimes so small that it is trivial to investigate it.

It should be noted that the definition of a set of reduction rules and the proof of their soundness only have to be performed once. This means that it is only the actual reduction and the investigation of the reduced CP-net which have to be carried out each time we want to investigate a new CP-net. The soundness proof is often complex and cumbersome, because we have to consider all possible reductions of all possible CP-nets. Notice, however, that a user can apply an existing set of reduction rules without having to know the soundness proof.

A serious problem for many reduction methods is the fact that the results obtained from them are non-constructive – in the sense that the absence of a property in the reduced net does not tell much about why the property is absent in the original net.

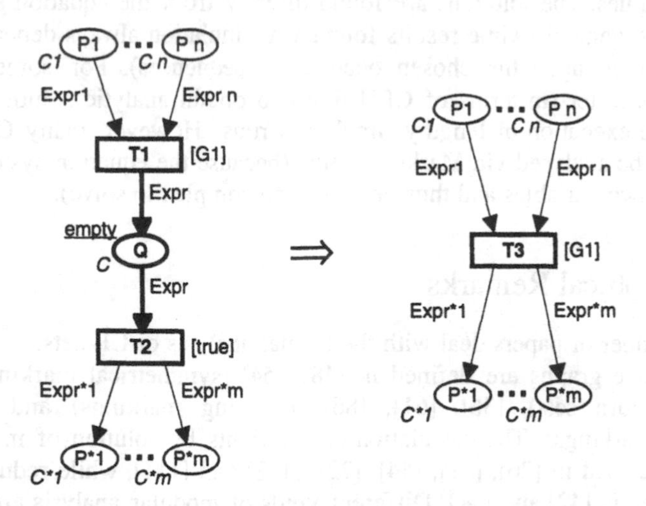

Fig. 5.3. Simple reduction rule preserving many kinds of dynamic properties

5.4 Performance Analysis

Most applications of CP-nets are used to investigate the logical correctness of a system. This means that we consider the dynamic properties of Chap. 4 and the functionality of the system (e.g., whether it produces the expected results). However, CP-nets can also be used to investigate the performance of a system (e.g., the maximal time used for the execution of certain operations and the mean waiting time for certain requests). To perform this kind of analysis, it is convenient to extend the CPN model with a time concept. We then specify how the different activities and states "consume" time.

Most time extensions add a **global clock** (which may be continuous or discrete), and this means that we now can talk about the time interval in which a marking exists. The time consumption can be described in many different ways. One possibility is to enforce a delay between the enabling and the occurrence of a transition. However, it is also possible to have a delay between the removal of input tokens and the creation of output tokens, or a delay between the creation of a token and the time at which that token can be used. In all three cases, the delay may depend upon the involved colours, and it may be specified as a constant value, a value chosen randomly inside a given interval, or a value determined by a probability distribution.

Performance analysis can be made via simulation, and in Chap. 6 we shall describe how the CPN simulator supports this kind of analysis. However, there also exist more formal approaches to performance analysis. For some kinds of delays, it is possible to translate the CP-net into a Markov chain (which is a well-known type of statistical model). The Markov chain determines an equation system, by means of which we can find analytic solutions to the different performance values. The solutions are found directly from the equation system and thus they are general (while results found via simulation always depend, at least to some extent, upon the chosen occurrence sequences). For some kinds of CP-nets it is faster (in terms of CPU time) to obtain analytic solutions – compared to the execution of lengthy simulation runs. However, many CP-nets are too large to be analysed via Markov chains (because the equation system has too many unknown variables and thus becomes too complex to solve).

Bibliographical Remarks

A large number of papers deal with the formal analysis of CP-nets.

Occurrence graphs are defined in [48], [54] (symmetrical markings), [113], [115] (stubborn sets), [36], [61], [86] (covering markings) and [19], [69] (symbolic markings). The calculation of invariants by solution of matrix equations is described in [26], [51], [54], [72], [102] and [111], while reduction rules are described in [42] and [44]. Different kinds of modular analysis are proposed in [18], [23], [79] and [114].

Performance analysis is one of the largest and most important subareas of Petri nets. Different time extensions of CP-nets are described in [17], [20], [31], [55], [57], [67] and in many of the papers contained in [89] and [90].

Exercises

Exercise 5.1.
Consider the resource allocation system from Sect. 1.2, and the full occurrence graph in Fig. 5.1.

(a) Check that the occurrence graph is correct (e.g., by using a CPN simulator).

(b) Use the occurrence graph to verify the upper and lower bounds which we have postulated at the end of Sect. 4.2.

(c) Use the occurrence graph to verify the home space which we have postulated at the end of Sect. 4.3.

(d) Use the occurrence graph to verify the liveness properties which we have postulated at the end of Sect. 4.4.

(e) Use the occurrence graph to verify the fairness properties which we have postulated at the end of Sect. 4.5.

Exercise 5.2.

Consider the process control system from Exercise 4.6. This exercise only makes sense if you have access to an occurrence graph tool. Even though the net is rather small, it will take too long to produce the occurrence graphs if this has to be done manually.

(a) Construct a full occurrence graph (for |PROC| = |RES| = 2).

(b) Construct an occurrence graph reduced by symmetrical markings (for |PROC| = |RES| = 2).

(c) Use the occurrence graphs of (a) and (b) to investigate the boundedness properties of the process control system.

(d) Use the occurrence graphs of (a) and (b) to investigate the home properties of the process control system.

(e) Use the occurrence graphs of (a) and (b) to investigate the liveness properties of the process control system.

(f) Use the occurrence graphs of (a) and (b) to investigate the fairness properties of the process control system.

(g) Repeat (a)–(f) for |PROC| = |RES| = 3.

Exercise 5.3.

Consider the resource allocation system from Sect. 1.2.

(a) Use the invariants from Sect. 5.2 to verify the lower bounds which we have postulated at the end of Sect. 4.2.

Exercise 5.4.
Consider the data base system from Sect. 1.3 – for which we postulate the invariants shown below. The function Rec maps each MES multi-set into the DBM multi-set which is obtained by replacing each MES-token with its second component (i.e., the receiver). The function Ign maps each DBM-token into the E multi-set which has the same number of tokens. Intuitively, this means that we ignore the token colour.

PI_{DBM}	Inactive+Waiting+Performing = DBM
PI_{MES}	Unused+Sent+Received+Acknowledged = MES
PI_{E}	Active+Passive = E
PI_{PER}	Performing = Rec(Received)
PI_{WA}	Mes(Waiting) = Sent+Received+Acknowledged
PI_{AC}	Ign(Waiting) = Active

(a) Use the invariants to verify the bounds which we have postulated at the end of Sect. 4.2.

(b) Use the invariants to verify the liveness properties which we have postulated at the end of Sect. 4.4.

Exercise 5.5.
Consider the reduction rule in Fig. 5.3.

(a) Use the rule to reduce the telephone system in Figs. 3.6–3.8.

(b) Use the rule to reduce the philosopher system in Exercise 1.6 (b).

(c) Impose some restrictions on the arc expression, Expr, and discuss to what extent the rule then preserves the dynamic properties defined in Chap. 4.

(d) Use the results of (b) and (c) to find dynamic properties for the philosopher system.

Chapter 6

Computer Tools for Coloured Petri Nets

The practical use of CP-nets, just like all other description techniques, is highly dependent upon the existence of adequate computer tools – helping the user to handle all the details of a large description. For CP-nets we need an editor supporting construction, syntax check, and modification of CP-nets, and we also need a number of analysis programs supporting the different analysis methods. The recent development of fast and cheap raster graphics gives us the opportunity to work directly with the graphical representation of CP-nets (and occurrence graphs).

This chapter describes the basic ideas behind a concrete set of CPN tools. However, the chapter may also be seen as a presentation of a set of design criteria and design ideas, which is relevant to all Petri net tools. We describe the CPN editor and CPN simulator, i.e., tools to construct, modify, syntax check and simulate CP-nets. Such tools have already existed for several years, and they have been used in a number of industrial projects – some of which will be presented in Chap. 7. We also describe tools supporting the formal analysis methods introduced in Chap. 5. Such tools are under development, and the first prototypes are now available.

The most important advantage of using computerized CPN tools is the possibility of obtaining *better results*. As an example, the CPN editor provides the user with a precision and quality which by far exceed the normal manual drawing capabilities of human beings. Analogously, computer support for complex analysis methods makes it possible to obtain results which could not have been achieved manually, because the calculations would be too error-prone.

A second advantage is the possibility of obtaining *faster results*. As an example, the CPN editor multiplies the speed by which modifications can be made. It is easy to change the size, form, position, and text of the individual elements of a CPN diagram – without having to redraw the entire diagram. It is also possible to construct new parts of a net by copying and modifying existing subnets. Analogously, analysis methods may be fully or partially automated. As an example, the manual construction of an occurrence graph is an extremely slow process, whereas it can be done on a computer in a few minutes (or hours).

A third advantage is the possibility of making *interactive presentations* of the analysis results. The CPN simulator makes it easy to trace the different occurrence sequences in a CP-net. Between each step, the user can (on the graphical representation of the CP-net) see the enabled transitions and choose between them to investigate different occurrence sequences. Analogously, it is possible to make an interactive investigation of a complex occurrence graph using a special-purpose search system.

A fourth advantage is the possibility of *hiding technical aspects* of the CP-net theory inside the tools. This allows the user to apply complicated analysis methods without having a detailed knowledge of the underlying mathematics. As an example, the user may apply a tool supporting a reduction technique without knowing the details of the individual reduction rules or the soundness proof.

For industrial applications the possibility of producing fast results of good quality is a necessary prerequisite for the entire use of CP-nets. It is also important to have tools and techniques which allow CP-nets to be used without a deep knowledge of Petri net theory.

Section 6.1 presents the CPN editor, while Sect. 6.2 describes the CPN simulator. Finally, Sect. 6.3 deals with CPN tools for formal analysis.

6.1 Editing of CP-nets

The CPN editor allows the user to construct, modify and syntax check hierarchical CP-nets. It is also possible to construct many other kinds of diagrams – in particular those which build upon the concept of a mathematical graph (i.e., contain nodes and arcs). The CPN tools are designed to work with large CPN diagrams. In practice, we often deal with diagrams which have 50–100 pages, each with 5–25 nodes and 10–50 arcs. All figures in this book are produced by means of the CPN editor.

The user works with a high-resolution raster graphical screen and a mouse. It is recommended, but not necessary, to have a large colour screen. The CPN diagram under construction can be seen in a number of windows – where it looks as close as possible to the final output (e.g., obtained by a laser printer). The editor is menu driven and has self-explanatory dialogue boxes (as known, e.g., from many Macintosh programs). The user moves and resizes the objects by direct manipulation, i.e., by means of a mouse (instead of typing coordinates and object identification numbers on the keyboard).

The CPN tools are implemented for two different kinds of operating systems. The tools run on Macintosh machines, and they also run on all machines with Unix and X-Windows. In this book we describe the Macintosh interface. The X-Windows interface looks slightly different, but the functionality is the same.

Windows, menu bar and status bar

The CPN editor supports hierarchical CP-nets and this means that each CPN diagram contains a number of pages. A typical screen image looks as shown in Fig. 6.1, which illustrates how the hierarchical telephone system from Sect. 3.2 may be displayed during an editing session.

At the top of the screen we have the **menu bar**. When the user presses one of the words in the menu bar (e.g., File or Align) a popup menu is displayed, and a command from this menu can be chosen. Each menu contains a set of commands which are closely related to each other. As an example, the file menu contains all commands which deal with file operations (e.g., commands to save and open diagrams). The align menu contains commands which allow the user to align objects with respect to each other (e.g., horizontally or vertically). Later in this section we shall give a brief overview of all the different menus. The apple symbol (at the left end of the menu bar) belongs to the general Macintosh interface (hence it is missing in the X-windows implementation).

Each page of the hierarchical CP-net is displayed in its own **window**, which may be open (i.e., visible) or closed. In Fig. 6.1 there are three open windows: *Phone#1*, *Dialing#3* and *BreakSen#5*. At the top of each window, there is a title bar which contains the identification of the page, i.e., the page name and page number. One window is fully visible, because it is displayed on top of the others; this window is active, while the others are passive. Along its border, the active window has a set of controls, which are invoked by pressing the mouse inside them. At the left end of the title bar there is a small rectangular box which can be used to close the window. At the right end there is a zoom box which is used

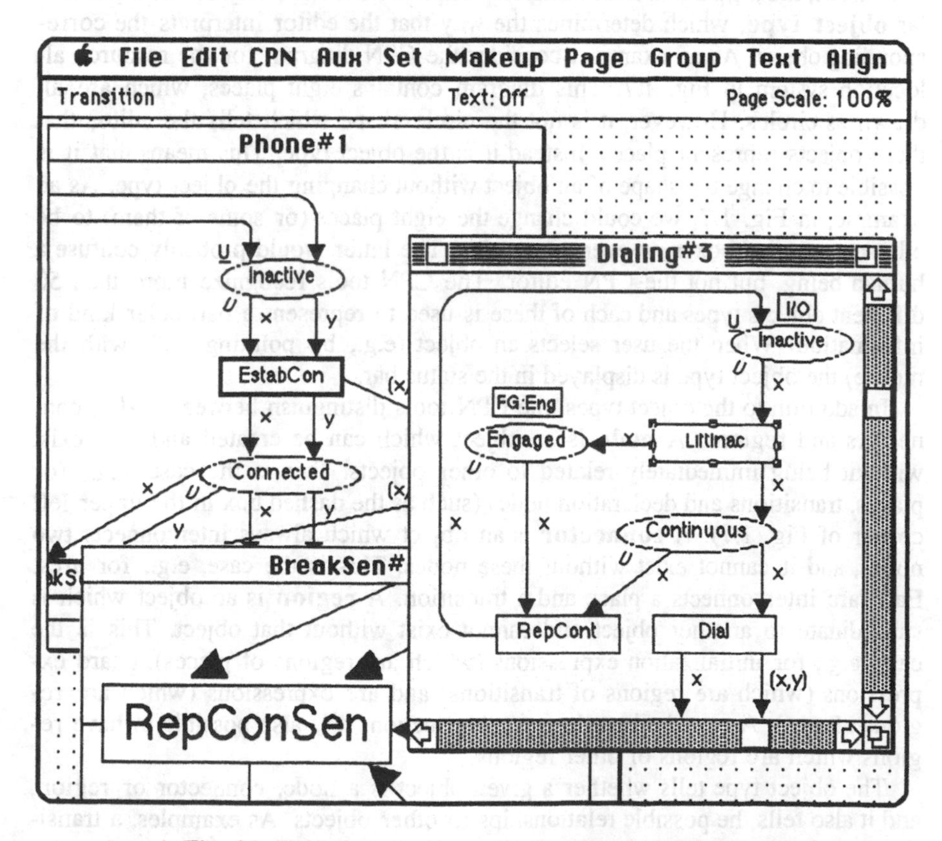

Fig. 6.1. Typical screen image displayed by the CPN editor

to change the size of the window, so that it covers the entire screen. At the right and bottom borders there are two scroll bars. They are used to change the view area. As an example, the user may press the up arrow (immediately below the zoom box). This will move the view area upwards (*Dial* and *RepCont* will disappear, while the area above *Inactive* becomes visible). The two small rectangles in the scroll bars indicate the current position of the view area. For the moment we see an area which is positioned at the upper right part of the page. Between the two arrows in the lower right corner we find the shape box. It is used to change the size of the window. When the mouse is pressed inside a passive window it becomes active. This means that it is drawn on top of the other windows (and gets visible controls).

Immediately below the menu bar is the **status bar.** It contains information about the present state of the CPN tool. In Fig. 6.1 we are told that the currently selected object *LiftInac@Dialing#3* is a transition. We are not in text mode and the active page *Dialing#3* is displayed with standard scale (i.e., 100%).

Different types of objects

The pages of a CPN diagram contain a number of graphical objects, e.g., places, transitions, arcs, guards and initialization expressions. Each object has a particular **object type**, which determines the way that the editor interprets the corresponding object. As an example, consider the CPN diagram for the resource allocation system in Fig. 1.7. This diagram contains eight places, which are all drawn as circles. However, it is not the circle shape which tells the editor that these objects represent places. Instead it is the object type. This means that it is possible to change the shape of an object without changing the object type. As an example, in Fig. 1.7, we could change the eight places (or some of them) to be ellipses, rounded boxes or even rectangles. The latter would probably confuse a human being, but not the CPN editor. The CPN tools recognize more than 50 different objects types and each of these is used to represent a particular kind of information. When the user selects an object (e.g., by pointing to it with the mouse) the object type is displayed in the status bar.

In addition to the object types, the CPN tools distinguish between nodes, connectors and regions. A **node** is an object which can be created and may exist without being immediately related to other objects. This is the case, e.g., for places, transitions and declaration nodes (such as the dashed box in the upper left corner of Fig. 1.7). A **connector** is an object which always interconnects two nodes, and it cannot exist without these nodes. This is the case, e.g., for arcs. Each arc interconnects a place and a transition. A **region** is an object which is subordinate to another object and cannot exist without that object. This is the case, e.g., for initialization expressions (which are regions of places), guard expressions (which are regions of transitions) and arc expressions (which are regions of arcs). As we shall see later in this section, it is also possible to have regions which are regions of other regions.

The object type tells whether a given object is a node, connector or region, and it also tells the possible relationships to other objects. As examples, a transition may have a guard region, but it cannot have an initialization region (because these can only be related to places). The transition may also have one or more arcs connecting it with other nodes (which have to be places).

The division of objects into nodes, connectors and regions makes it possible for the CPN editor to recognize a CPN diagram as a mathematical graph (and not just as a set of unrelated graphical objects). When the user constructs a connector, he simply identifies the two nodes which he wants to interconnect (by pointing somewhere inside them). Then the system draws the connector and it automatically calculates the position of the connector end points, in such a way that they are precisely on the node borders. This saves time and it makes the produced diagram nicer (because the editor can position the end points more precisely than the user). It is also important to notice that the editor automatically redraws the connectors and the regions each time this becomes necessary, for example because a node is repositioned, resized or deleted. The effects of these operations are shown in Fig. 6.2. A repositioning implies that the regions keep their relative position (with respect to the node). A resizing implies that the relative positions of the regions are scaled while their sizes are either unchanged or scaled (depending upon an attribute of each region). When a node is deleted, the regions and arcs are deleted too. Similar rules apply for the repositioning, resizing and deletion of arcs and regions.

The automatic redrawing described above may seem obvious, but it should be noticed that it only is possible because the CPN editor recognizes the structure of a CPN diagram as a mathematical graph. In most general-purpose drawing tools (e.g., MacDraw or MacDraft) a CP-net would only be recognized as a set of unrelated objects. This means that the repositioning or resizing of a transition has to be followed by a tedious manual repositioning of the guard, the end points of the arcs, and the arc expressions. Analogously, the deletion of a transition has to be followed by a manual deletion of the guard, the arcs and the arc expressions (and if this is forgotten we get an inconsistent net which, e.g., may have dangling arcs).

The CPN editor also distinguishes between CPN objects, auxiliary objects and system objects. Nearly all objects which we have met so far are **CPN objects**. This means that they have a formal meaning and they influence the behaviour of the CP-net. Places, transitions, arcs, initialization expressions, guards and arc expressions are examples of CPN objects. However, it is also useful to be able to have **auxiliary objects** in a CPN diagram. Such objects have no formal mean-

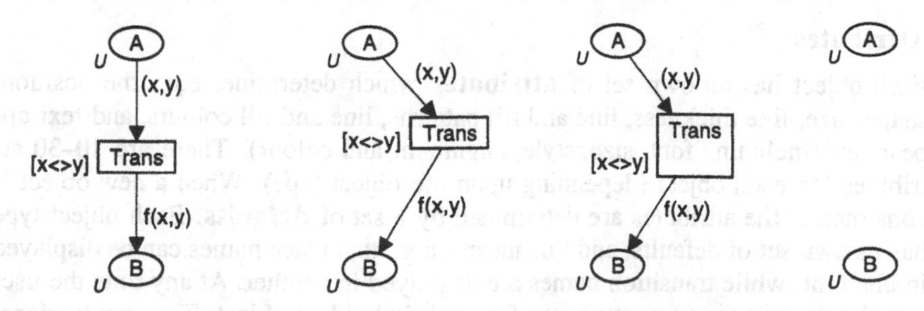

Fig. 6.2. When a node is repositioned, resized or deleted, the regions and surrounding arcs are automatically redrawn

ing. They are included to make the CP-net more readable, and thus they play a similar role to comments in programming languages. It is possible to create auxiliary objects of many different shapes, e.g., ellipses, rectangles, rounded boxes, polygons, wedges and bit maps. The auxiliary objects may be nodes, connectors or regions, and there are no restrictions on the relationship to other objects. As an example it is possible to create an auxiliary connector between any pair of nodes (e.g., between two places, or between an auxiliary node and a declaration node). All objects may have auxiliary regions. Finally, there are **system objects** which are special objects created by the CPN editor itself. An example is the current marking regions which the CPN simulator creates for each place. These regions contain information about the number and the colour of the tokens which reside on the corresponding place, in the current marking (cf. Fig. 1.7). Other examples of system objects are the nodes, connectors and regions which constitute the page hierarchy graph (cf. Figs. 3.4 and 3.8).

Graphical representation

One of the most attractive features of CP-nets (and Petri nets in general) is the existence of a nice graphical representation. It would thus be a pity to put narrow restrictions on this representation, for example by making an editor in which all places and transitions have a fixed shape and size. In our opinion, a good editor must allow the user to draw almost all kinds of CP-nets which can be constructed by a pen and a typewriter.

In the CPN tools it is possible for the user to determine, in great detail, how he wants the CPN diagrams (and the occurrence graphs) to look. This means that the modeller for each object can determine the position and the detailed appearance, e.g., the shape. Connectors can be single headed, double headed or without heads. Nodes and regions can be boxes, rounded boxes, ellipses, polygons, wedges, pictures and labels. A picture is a bit map. It can be a snapshot of a part of a CPN page or it can, via the clipboard, be imported from other programs, e.g., MacPaint. Pictures make it easy to work with icons. As an example, a substitution transition may be a picture, which is a diminished snapshot of the corresponding subpage. A label is a rectangular object where the size is automatically determined from the extent of the text. Usually, we use labels for most of the CPN regions.

Attributes

Each object has its own set of **attributes** which determine, e.g., the position, shape, size, line thickness, line and fill patterns, line and fill colours, and text appearance (including font, size, style, alignment and colour). There are 10–30 attributes for each object (depending upon the object type). When a new object is constructed, the attributes are determined by a set of **defaults**. Each object type has its own set of defaults, and this means, e.g., that place names can be displayed in one font, while transition names are displayed in another. At any time the user can change one or more attributes for each individual object. This can be done, e.g., by means of the six attribute commands. They divide the attributes into text, graphical, shape, region, page and mode attributes, respectively. The first three kinds of attributes are defined for all kinds of graphical objects, while region attributes only are defined for regions and page attributes for pages. Mode attributes are defined for pages and for substitution transitions. They do not influence the graphical layout. Instead they determine the set of prime pages, and they specify the parts (of the entire model) which are going to participate in the simulation/analysis.

The dialogue box for region attributes is shown in Fig. 6.3. It allows the user to specify where he wants the region to be positioned (by determining the desired offset from the parent object). Moreover, it can be specified whether the size of the region should change when the parent is resized, and whether the region always has the same colour as the parent. Finally, it is specified whether a popup region (of the region in question) is initially shown, hidden or missing. This part is only present in the dialogue box when the selected region is a key region. We shall return to key and popup regions later in this section. The *Save* button of the dialogue box allows the user to save the specified attribute values as new defaults to apply to new objects of the corresponding object type. When defaults are saved the user determines whether he wants the new defaults to become diagram defaults (i.e., be in effect only for the present diagram), to become system defaults (i.e., be in effect for all new diagrams), or both. The *Load* button allows the user to copy the values of the existing diagram or system defaults into the dialogue box.

In one of the next versions of the CPN tools it will also be possible to copy each of the six kinds of attributes from one object to another. This is much faster than invoking the corresponding dialogue box. Finally, it is possible to create a new object by copying an existing object. Then the new object automatically gets the same attributes as the old object.

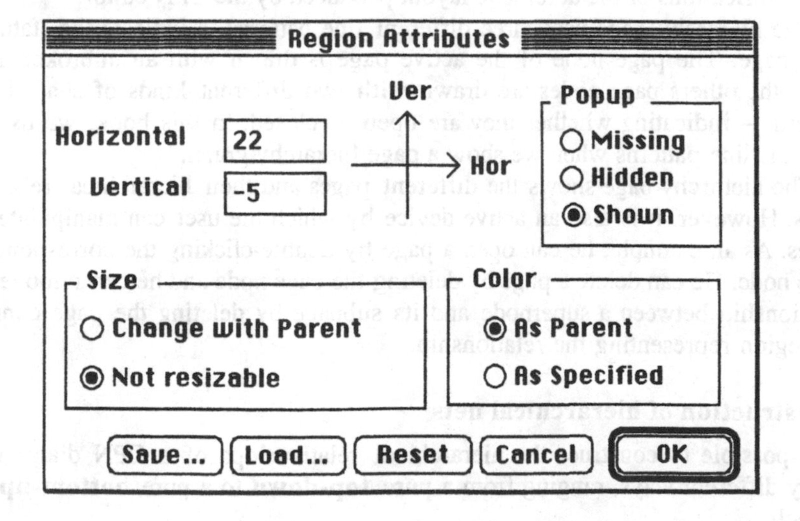

Fig. 6.3. Dialogue box for the Region Attributes command

Options

In addition to the attributes, the CPN tools have a large set of **options** which determine how the detailed operations of the tools are performed. An attribute relates to a particular object, while an option relates to the entire CPN diagram. For example there is an option which determines how fast the contents of windows scroll. Another option specifies how duplicate arcs are treated when one or more nodes are merged into a single node (by means of the Merge command). Options are divided into a number of related classes, in a similar way to attributes. There are twelve such classes and each of them has its own Option command. For options we only have system defaults and no diagram defaults (because a diagram only has one value for each option). Option **defaults** can be saved and loaded in a way which is similar to attribute defaults.

Hierarchy page

The hierarchical relationships between the individual pages of a CPN diagram are shown in a **page hierarchy** graph, such as Figs. 3.4 and 3.8. The page hierarchy graph is displayed on a separate page called the **hierarchy page** and this page is automatically updated by the CPN editor – as the user creates and deletes pages (and hierarchical relationships between them). All the objects of the page hierarchy graph are **page objects**, which are a particular class of system objects. Like all other object types, page objects have their own sets of defaults (which determine how they look). Moreover, there is a class of options, called hierarchy options, which determines the standard layout of the hierarchy page, e.g., the spacing between the page nodes. Page objects can be moved and modified, in exactly the same way as all other types of objects. This means that the user can determine in great detail how the page hierarchy looks. For an example, consult the hierarchy page in Fig. 7.1, which is obtained by making a few manual modifications of the automatic layout produced by the CPN editor.

The hierarchy page uses three different line patterns to indicate the status of each page. The page node of the active page is drawn with an unbroken line, while the others page nodes are drawn with two different kinds of shaded line patterns – indicating whether they are open or closed. In this book, we usually omit the line patterns when we show a page hierarchy graph.

The hierarchy page shows the different pages and their hierarchical relationships. However, it is also an active device by which the user can manipulate the pages. As an example, he can open a page by double-clicking the corresponding page node. He can delete a page by deleting the page node and he can remove the relationship between a supernode and its subpage by deleting the page connector/region representing the relationship.

Construction of hierarchical nets

It is possible to construct the hierarchical relationships of a CPN diagram in many different ways, ranging from a pure **top-down** to a pure **bottom-up** approach.

When a page gets too many places and transitions, we can move some of them to a new subpage. This is done by a single editor operation. The user selects the nodes to be moved and invokes the Move to Subpage command. Then the editor

- checks the legality of the selection (it must form a closed subnet),
- creates the new page,
- moves the subnet to the new page,
- creates the port places by copying those places which were next to the selected subnet,
- calculates the port types,
- creates the corresponding port regions,
- constructs the necessary arcs between the port nodes and the selected subnet,
- prompts the user to create a new transition which becomes the supernode for the new subpage,
- draws the arcs surrounding the new transition,
- creates a hierarchy inscription for the new transition,
- updates the hierarchy page.

The effect of such an operation is illustrated by Figs. 6.4 and 6.5, which show the data base system before and after such a command. The grey rectangles around the four rightmost nodes of Fig. 6.4 indicate that these nodes are selected.

As may be seen, a lot of rather complex checks, calculations and manipulations are involved in the Move to Subpage command. However, almost all of these are automatically performed by the CPN editor. The user only selects the subnet, invokes the command and creates the new supernode. The rest of the work is done by the CPN editor. This is of course only possible because the CPN editor recognizes a CPN diagram as a hierarchical CP-net, and not just as a mathematical graph or as a set of unrelated objects. Without this property the user would have to do all the work by means of the ordinary editing operations (which allow him to copy, move and create the necessary objects). This would be possible – but it would be much slower and much more error-prone.

In Fig. 6.5 we have three port assignments (created by the Move to Subpage command):

- *Sent@DataBase#1 —> Sent@New#2,*
- *Inactive@DataBase#1 —> Inactive@New#2,*
- *Acknowledged@DataBase#1 —> Acknowledged@New#2.*

For all three assignments, the name of the port node is identical to the name of the socket node. Such assignments are (by convention) not listed in the hierarchy inscription. However, if the user at *DataBase#1* changes the name of *Sent* to *Mailed*, the hierarchy inscription gets an additional line "Mailed —> Sent".

There is also an editor command which turns an existing transition into a supernode – by relating it to an existing page. Again, most of the work is done by the editor. The user selects the transition and invokes the command. Then the editor:

- makes the hierarchy page active,

- prompts the user to select the desired subpage; when the mouse is moved over a page node it blinks, unless it is illegal (because selecting it would make the page hierarchy cyclic),
- waits until a blinking page node has been selected,
- tries to deduce the port assignment by means of a set of rules which looks at the port/socket names and the port/socket types,
- creates the hierarchy inscription with the name and number of the subpage and with those parts of the port assignment which could be automatically deduced,
- updates the hierarchy page.

Finally, there is an editor command which replaces a supernode by the entire content of its subpage. Also, this operation involves a lot of complex calculations and manipulations, but again all of them are done by the CPN editor. The user simply selects the supernode, invokes the command and uses a simple dialogue box to specify the details of the operation (e.g., whether the subpage shall be deleted when no other supernode uses it).

The three hierarchy commands described above can be invoked in any order. A user with a top-down approach would typically start by creating a page where each transition represents a rather complex activity. Then a subpage is created for each activity. The easiest way to do this is to use the Move to Subpage command. Then the subpage automatically gets the correct port places, i.e., the correct interface to the supernode. As the new subpages are modified, by adding places and transitions, the subpages may become so detailed that additional levels of subpages must be added. This is done in exactly the same way as the first level was created.

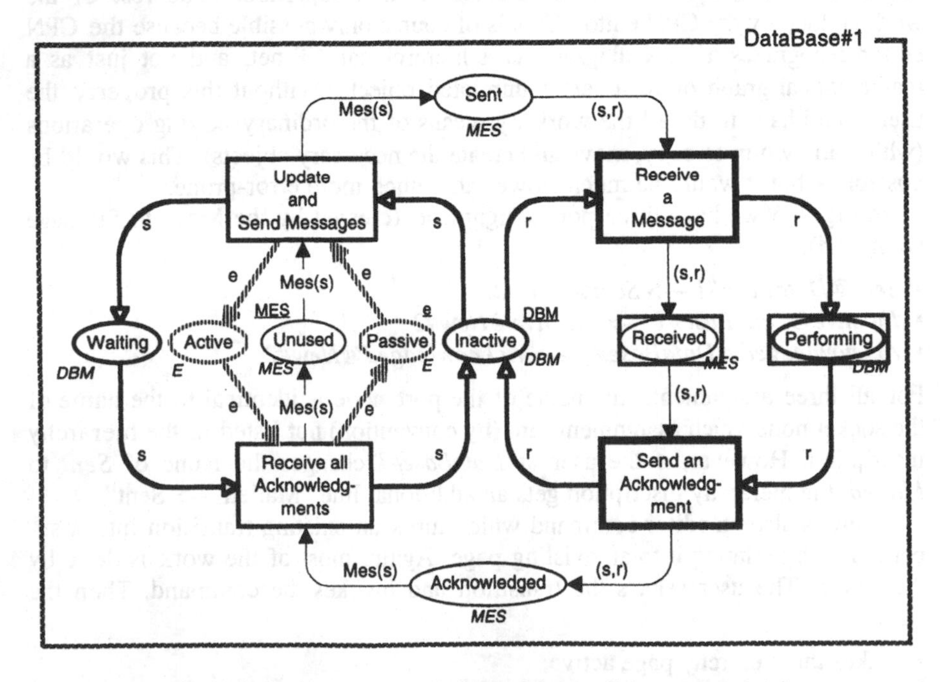

Fig. 6.4. Data base system before a Move to Subpage command

In contrast to this, a user with a bottom-up approach would start by creating a number of pages with the basic components of the modelled system. Later he would model the more abstract layers and relate these to the existing pages by means of the command that turns transitions into supernodes. In practice, it is

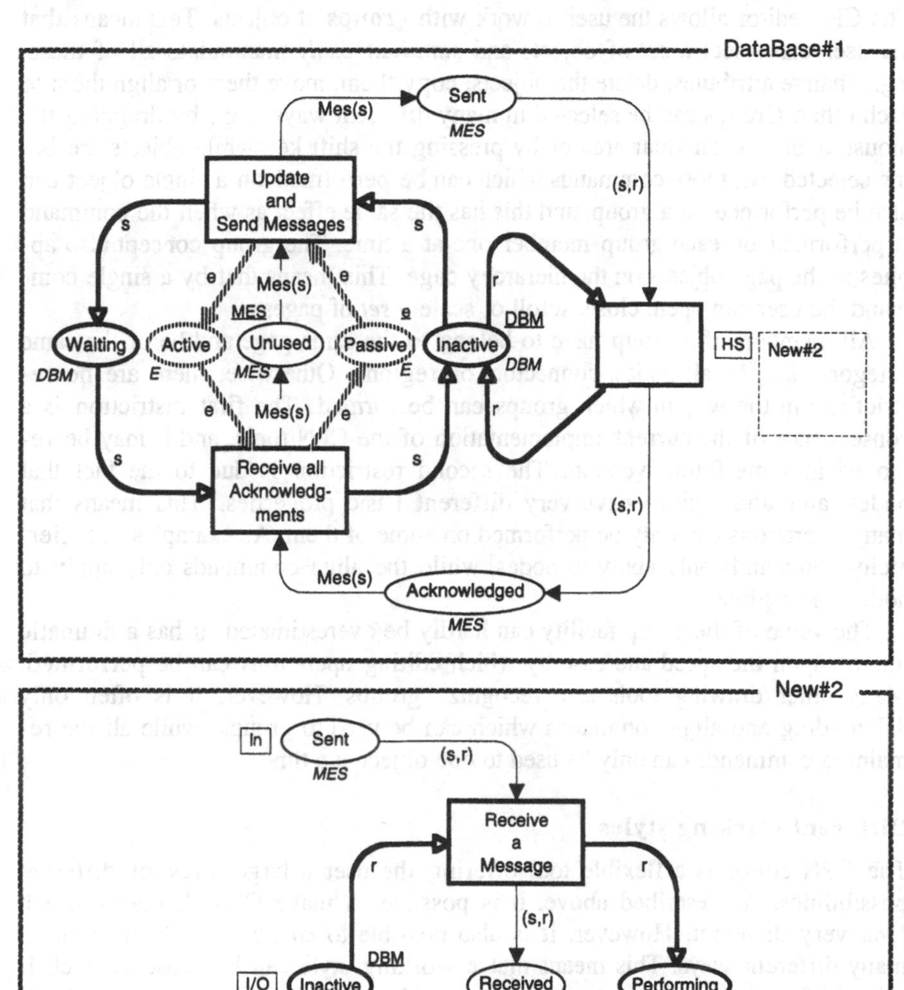

Fig. 6.5. Data base system after a Move to Subpage command

very rare to find pure top-down or bottom-up modelling. Most users alternate between the two strategies.

Groups of objects

The CPN editor allows the user to work with **groups** of objects. This means that the user can select a *set* of objects and *simultaneously* manipulate all of these, e.g., change attributes, delete the objects, copy them, move them or align them to each other. Groups can be selected in many different ways, e.g., by dragging the mouse over a rectangular area or by pressing the shift key while objects are being selected. All those commands which can be performed on a single object can also be performed on a group, and this has the same effect as when the command is performed on each group member one at a time. The group concept also applies to the page objects on the hierarchy page. This means that by a single command the user can open, close, scroll or scale a *set* of pages.

All members of a group have to belong to the same page and be of the same category, i.e., be all nodes, connectors or regions. Otherwise, there are no restrictions in the way in which groups can be formed. The first restriction is a consequence of the current implementation of the CPN tools, and it may be removed in some future version. The second restriction is due to the fact that nodes, arcs and regions have very different basic properties. This means that many operations can only be performed on some of them. As examples, the hierarchy commands only apply to nodes, while the align commands only apply to nodes and regions.

The value of the group facility can hardly be overestimated. It has a dramatic impact upon the speed and ease by which editing operations can be performed. Many other drawing tools also recognize groups. However, it is often only delete, drag and align commands which can be used to groups, while all the remaining commands can only be used to one object at a time.

Different working styles

The CPN editor is a flexible tool offering the user a large array of different possibilities. As described above, it is possible to make CPN diagrams which *look* very different. However, it is also possible to *construct* each diagram in many different ways. This means that a working style can be chosen, which is suitable for the needs and temperament of the modeller. One example of this flexibility is the many different ways in which the hierarchical relationships between pages can be constructed.

Another example is the fact that the CPN editor allows the user to construct the individual objects of a CP-net in many different ways. Some users prefer to make the net structure first, i.e., the places, transitions and arcs. Later they then add the net inscriptions, i.e., the CPN regions. They either finish one node at a time or finish one kind of CPN regions at a time. They either type from scratch or they copy text from existing regions. Other users prefer to create templates (e.g., a transition with a guard region and a place with a colour set region and an initialization region). Then the diagram is created by copying the templates until the appropriate number of places, transitions and arcs have been made. During this process the new objects are positioned, and if necessary their inscriptions are modified. When a node is being copied, the editor automatically includes the regions and for a group of nodes also the internal connectors.

Most users prefer to work in a way which is a mixture of those described above. Thus it is extremely important to support as many different working styles as possible – so that they can all be performed in a natural and efficient way. Hence the CPN editor has been designed to allow most operations to be performed in several different ways.

Complex diagrams

A CPN diagram contains many different kinds of information and this means that the individual pages very easily become cluttered. To avoid this, the user is allowed to make objects invisible. As an example, the user may hide all colour set regions and instead indicate the different colour sets by giving the corresponding places different graphical attributes as shown in Fig. 1.35 and Fig. 3.9. It should be noted that the CPN tools interpret the invisible regions in the same way as all other regions. It is the regions that determine the semantics – not the encoded graphical attributes. The user must keep the graphical attributes updated when he changes the invisible regions (and vice versa). Otherwise, he may easily be confused.

Another facility, to avoid cluttered diagrams, is the concept of key and popup regions. It is used for a number of different object types – both in the editor and in the simulator. The idea is very simple. Instead of a single region with a lot of information, we have both a key region and a popup region. The **key region** is a region of the object to which we want to attach the information. The key region is small. Usually it only contains one or two characters. Its main purpose is to give access to the **popup region** which contains the actual information. The popup region is a region of the key region, and this means that a repositioning of the key also repositions the popup. A double-click on the key region toggles the visibility of the popup region. Hence it is extremely easy to hide and show large amounts of information. Key and popup regions are used for hierarchy inscriptions. The key contains the letters HS, while the popup contains the identity of the subpage and the port assignment (which may be quite large). In the CPN simulator, key and popup regions are used to display the current marking of each place. Here the key contains the number of tokens, while the popup contains a textual representation of the actual colour values. The latter may be very large. There are applications in which some tokens have a colour which is a list of 50,000 records.

The use of key and popup regions is in many respects similar to the use of popup windows. The difference is that the popup regions are objects of the diagram itself. Thus they can be handled in exactly the same way as all other objects, e.g., copied. An attribute of each key region records whether the corresponding popup region is shown, hidden or missing. The diagram default of this attribute determines the initial state of the popup region. A missing popup region can always be created, with the correct information. This is done simply by double-clicking the key.

Use with care

The generality of the CPN editor means that it is possible to create very confusing CPN diagrams. As examples, it may be impossible to distinguish between auxiliary objects and CPN objects (because they have been given identical attributes), transitions may be drawn as ellipses while places are boxes, and some or all of the objects may be invisible – just to mention a few possibilities. We do not think it makes sense to construct a tool which makes it *impossible* to produce bad nets. Such a tool will, in our opinion, inevitably be far too rigid and inflexible. However, we do think that a good tool should make it easy for the user to make good nets and easy to avoid tricky constructions, unless these are really wanted.

In the next subsections we give a brief discussion of some of the other facilities in the CPN editor. This is done by looking at one menu at a time.

File menu

A command allows the user to create new diagrams. Moreover, he can open, close, save and print existing diagrams. It is also possible to save and load individual pages. This means that a page can be moved from one diagram to another. In a later version, it will be possible to save and load subdiagrams (i.e., groups of pages) without destroying the hierarchical relationships between them. This will make it easy to have libraries of reusable submodels. The contents of text files can be loaded into the texts of nodes, and vice versa. Finally, there is a command to import diagrams which are created by other tools (e.g., the IDEF_{CPN} diagrams described in Sect. 7.3).

Edit menu

This menu contains the standard undo, redo, cut, copy, paste and clear commands known from the Macintosh concept. For the moment, undo and redo only work for a limited set of commands. There is also a command to get detailed information about an individual object, e.g., how many regions it has.

CPN menu

We have already described the commands to create and destroy hierarchical relationships. In addition to these, there are commands to create places, transitions, arcs, CPN regions and declaration nodes. There are also commands to define fusion sets, specify port nodes and perform port assignments. Finally, there is a command to start a syntax check.

Aux menu

This menu contains commands to create auxiliary objects of all different shapes. There are also commands to turn auxiliary objects into CPN objects, and vice versa. Finally, there are commands to start and stop the SML compiler.

Set menu

This menu contains all the commands to change attributes, options and the defaults of attributes and options.

Makeup menu

A command allows the user to hide some objects, temporarily, while another object is being selected. This is useful when many objects are close to each other or on top of each other. Objects can be dragged to a new position on the same page or moved to another page. The size of objects can be adjusted, to an arbitrary size or to a size that fits the extent of the text it contains. The shape of objects can be changed, e.g., from a rectangle to an ellipse or from a rounded box to a picture. Nodes can be merged into other nodes and they can be duplicated. Each new node gets a set of regions and a set of surrounding arcs which are similar to those of the original node. There are commands to hide and show regions and to change the graphical layering of objects. Finally, there are commands to jump to the parent object, oldest child object, next object and previous object (in the region and layering hierarchy). This can also be done by means of the arrow keys.

Page menu

We have already mentioned the commands to open, close, scroll and scale pages. In addition, there is a command to create new pages and a command to redraw the page hierarchy graph. The latter is used when the hierarchy page becomes too cluttered, e.g., because the user has made a number of manual changes to the automatic layout maintained by the CPN editor.

Group menu

This menu contains commands to select different kinds of groups, e.g., all nodes on a given page, all connectors, all regions, or all members of a given fusion set. In a later version, there will be added more elaborated group commands. As an example, it will then be easy to select all the places which have a given colour set. This is useful when colour sets are made invisible and replaced by a graphical encoding of their contents.

Text menu

This menu contains all the commands to manipulate text. There are commands to search for specified text strings and replace them by others – either in the entire diagram, on a single page, or in the selected objects. It is also possible to scroll to the beginning of a large text and to search for matching brackets. In a later version, commands will be added that make it possible to create hypertexts, i.e., texts with pointers to other texts. For the moment hypertexts are only used in connection with the error reporting after a syntax check.

Align menu

The commands in this menu make it easy to align nodes and regions to each other. There are many different possibilities. The objects may, e.g., be positioned vertically below each other, or in such a way that the upper edges are horizontally aligned. Objects may also be positioned with equal distances between them – on a line or on a circle.

Consistent user interface

To make it easier to use the CPN tools we have tried to make the user interface as consistent and self-explanatory as possible. For this purpose, we have defined a set of concepts which allow a precise description of the different parts of the interface. As an example, we have defined a set of standard buttons to be used in dialogue boxes. Some of these are concepts from the general Macintosh interface, e.g., *Cancel, Reset* and *OK*. Others are special for the CPN tools, e.g., *Save* and *Load* (of default values).

Another example is the use of list boxes, such as the one shown in Fig. 6.6. Depending upon the content and the purpose of the list box, the CPN tools will allow the user to select only a single line, a contiguous set of lines, an arbitrary set of lines, or no lines at all. When the dialogue box is opened, the list box may get the same selection as the previous time, have the first line selected, have no line selected, or have a selection which depends upon the current selection in the diagram or the content of the file system. By identifying the different ways in which a list box can behave, we have made it possible to give a precise description of the individual commands, and we have obtained a more consistent user interface.

Different skill levels

The CPN editor can be used at many different skill levels. Casual and novice users only have to learn and apply a rather small subset of the total facilities. The more frequent and experienced users gradually learn how to use the editor more efficiently. All the more commonly used commands can be invoked by means of

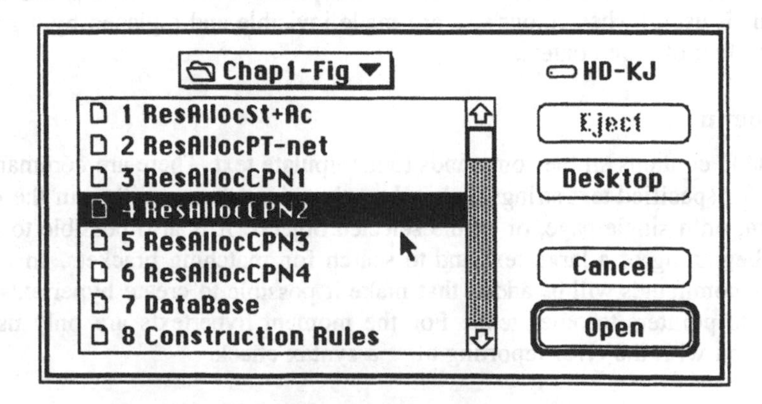

Fig. 6.6. List box

key shortcuts, and these can be changed by the user. Many commands have one or more **modifier keys** allowing the user to perform an operation, which otherwise would require several commands.

The user can create a **palette page** from which different objects can be copied as they are needed. The user simply clicks at the desired object, and then at all the positions where he wants a copy of the object. Palette pages are created and modified like all other pages. They simply have an attribute telling that they are palettes. This means that during the modelling, the user can create new palette pages and modify existing palettes. He can also move, size and scale the palettes as he wants. Palettes are useful, for example, when a company or a group of researchers want to impose a common standard for the graphics of their CPN diagrams.

Standard ML

The CPN tools use the CPN ML language for declarations and net inscriptions. This language is based on a well-known functional programming language, called Standard ML (SML).

Using an existing language gives us three different advantages. First of all, we have got a better, more general and more tested language, than we could have hoped to develop ourselves. The creation of a new programming language is a very slow and expensive process. Secondly, we only had to port the compiler to the relevant kind of operating systems and integrate it with the CPN tools – instead of developing a new compiler from scratch. Thirdly, we are able to reuse the considerable amount of documentation and tutorial material which already exists for SML and for functional languages in general.

Why did we then choose SML? First of all, CPN uses types, functions, operations, variables and expressions in a similar way as a typed functional language. Hence it is convenient to build upon such a language. Secondly, we needed a language with a flexible and extendible syntax. This allows the user to write the declarations and net inscriptions in a way which is close to standard mathematics. As an example, we use + to denote multi-set addition. This may seem trivial, but in most languages it would not be possible. The + operator is *infixed*. This means that it is written between the two arguments. Moreover, + is *polymorphic*. This means that it works for multi-sets over all different types. Finally, + is *overloaded*. This means that the same operator symbol is also used for other operations, e.g., addition of integers and addition of reals.

SML is only one out of a number of languages which fulfil the requirements listed above. SML was chosen because it is one of the best known of these languages. Moreover, SML had commercially available compilers and some of the CPN tool developers already had good experiences with it. The choice of SML has never been regretted. We have often been amazed by the high quality and the generality of the language and the ease with which complex programs can be made. Thus we consider the choice of SML as one of the most successful design decisions in the CPN tool project. We have got a very powerful and general inscription language, and we have saved a lot of implementation time. SML also makes it easy to make a smooth integration between code segments and the net inscriptions. A code segment is a sequential piece of code attached to a transition. The code segment is executed each time the transition occurs, and it may, e.g., update files or do other forms of reporting. We shall return to code segments when we describe the CPN simulator.

The graphical parts of the CPN tools are written in C, but much of the more intrinsic code is written in SML. One example is the code that calculates the set of enabled bindings. This code is complex. It defines a function which takes a set of arc expressions and a guard as arguments. For each such set the function produces a new function, which is able to map a set of place markings into a set of enabled bindings. Using the pattern matching concept of SML, it was possible to implement the binding function in a few weeks. If we had used C, or a similar procedural language, the programming task would have been much more difficult.

Another example of the versatility of SML is our implementation of multi-sets. This was done by defining a polymorphic type constructor ms. It maps an arbitrary type S into a new type, which contains all elements of S_{MS} and is denoted by S ms. In addition to the ms constructor, we have declared a large number of polymorphic and sometimes overloaded functions and operations. These are used to manipulate multi-sets, for example, via the multi-set operations defined in Sect. 2.1.

The current CPN tools use two different SML compilers. The Macintosh implementation uses a compiler developed at Edinburgh University, while the Unix and X-windows implementation uses a more modern compiler developed by AT&T. It is possible to run the CPN tools and the SML compiler on two separate machines, connected via a local area network.

CPN ML

CPN ML is obtained by extending SML in three different ways. The first two extensions make the language easier to use – in particular for people who are not familiar with all the details of SML. The third extension is made to be able to define the scope of reference variables.

The first extension adds syntactic sugar for colour set declarations. This makes it very easy to declare the most common kinds of colour sets. It also means that the user can include a large number of predefined functions, just by mentioning their name in a colour set declaration. This is illustrated by the declare clauses explained at the end of Sect. 1.3. As examples of predefined functions, each enumeration type has a function mapping colours into ordinal numbers, each product type has a function mapping a set of multi-sets into their product multi-set, and each union type has a set of functions performing membership tests. We have chosen only to include those predefined functions which appear in a declare clause. This saves a lot of space in the ML heap, because most CPN diagrams only use few of the predefined functions. A later version of the CPN tools will automatically detect whether a given predefined function is used or not. Then it will no longer be necessary to have the declare clauses.

In addition to the standard colour set declarations introduced in Sect. 1.4, it is possible to use an arbitrary SML type as a colour set – as long as the standard equality operator exists and the type is non-polymorphic. Hence it is possible to declare abstract data types and turn them into colour sets. When this is done, the user has to define three extra functions. The first defines the order of the colours. The second represents each colour as a text string, and the third draws a random colour. The functions are used during a simulation, e.g., to display the current markings. For the standard colour sets the functions are automatically defined, but for a user defined type this is not possible.

The second extension of SML allows the user to introduce typed variables, as a part of the declarations. SML does not have variable declarations. Instead a value may be bound to a name, and this determines the current type of the name. Later the name may get a new value, and then it gets a new type. It would be possible to use the same strategy for CP-nets. However, we have found it more fruitful to demand that the user explicitly declares the type of each variable. This makes it possible to perform an exhaustive type checking.

The third extension allows the user to declare **reference variables** – with three different kinds of scope. Reference variables are non-functional elements of SML, and we only allow them to be used in code segments. We distinguish between global, page and instance reference variables – in the same way that we distinguish between global, page and instance fusion sets. A global reference variable can be used by all code segments in the entire CPN diagram, while a page and instance reference variable can be used only by the code segments on a single page. A page reference variable is shared by all instances of the page, while an instance reference variable has a separate value for each page instance.

SML (and hence CPN ML) is a syntactically sugared version of typed lambda calculus. This means that it is possible to define all kinds of mathematical functions – as long as they are computable. Notice that the use of SML means that the CPN tools accept a very general class of CPN diagrams. The user can declare arbitrarily complex functions and operations. As an example, many CPN diagrams make use of recursive functions defined on list structures.

Syntax restrictions

The CPN editor is syntax directed. This means that it recognizes the structure of CP-nets and automatically prevents the user from making many kinds of syntax errors. This is done by means of a large number of **built-in** syntax restrictions. All the built-in restrictions deal with the net structure and the hierarchical relationships. As an example, it is impossible to make an arc between two transitions, or between two places. It is impossible to give a transition a colour set region, or give a place two colour set regions. It is also impossible to create cycles in the substitution hierarchy, and to make an illegal port assignment, involving nodes which are not sockets or ports, or are positioned on a wrong page.

The CPN editor also operates with **compulsory** syntax restrictions. These restrictions are necessary in order to guarantee that a CPN diagram has a well-defined semantics – and hence they must be fulfilled before a simulation (or another kind of behavioural analysis) is performed. Many of the compulsory restrictions deal with the net inscriptions and thus with CPN ML. As an example, it is checked that each colour set region contains the name of a declared colour set

S, and that all the surrounding arc expressions have a type which is either S or S_{MS}. It is checked that the members of a fusion set all have the same colour set and equivalent initialization expressions. It is also checked that all identifiers in arc expressions and guards are declared, e.g., as variables or functions. Many of the compulsory syntax restrictions could have been implemented as built-in restrictions. However, this would have put severe limits on the way in which a user can construct and edit a CPN diagram. As an example, we could have demanded that each place always has a colour set. This would, however, imply that the colour set has to be specified at the moment the place is created. We could also have demanded that each arc expression is always of the correct type. This would, however, imply that a colour set cannot be changed without simultaneously changing all the surrounding arc expressions.

A later version of the CPN editor will also operate with **optional** syntax restrictions. These are restrictions which the user may choose to impose upon himself. It is often the case that a particular CPN model does not exploit all the possibilities offered by the CPN syntax. Then it may be useful to specify those facilities which we do not expect to use. This will allow the CPN editor to warn the user if one of the excluded facilities is used, nevertheless. The user can then investigate whether the use was intended – or due to an error. As an example, it can be specified that all port assignments are supposed to be injective, surjective and total functions. It can be specified that all arcs are supposed to have an explicit arc expression. Otherwise, a missing arc expression is a shorthand for the empty multi-set. It can also be specified that all place and transition names are supposed to be unique, in the entire diagram or on each page.

Syntax checking

The possibility of performing an automatic syntax check means that the user has a much better chance of getting a consistent and error-free CPN diagram. This is very useful – also in situations where the user is not interested in performing a simulation or other kinds of machine assisted analysis.

The detailed checking of the expression syntax and the type consistency is performed by the SML compiler. When an error is found, an error message is produced. The first line contains a short description of the error, while the remaining lines are an exact copy of the error message produced by the SML compiler. To see how a typical error looks, let us consider the resource allocation system of Fig. 1.7, and let us assume that we change the arc expression between A and T1 from (x,i) to x. This will result in an **error message**, which in the Macintosh implementation looks as shown below. In the Unix and X-windows implementation the error message is produced by another SML compiler, and hence it looks slightly different.

```
C.11 Arc Expression must be legal
Type clash in:  x : (P  ms)
Looking for a: P ms
I have found a: U
««135»»
```

C11 means that it is the 11th kind of compulsory restriction, while «««135»»» is a hypertext pointer – which allows the user to jump to the error position (i.e., to the arc with the erroneous arc expression). To perform a hypertext jump, the user positions the cursor inside the hypertext pointer and presses an arrow key.

To speed up the syntax check the CPN editor avoids duplicate tests. As an example, the same arc expression may appear at several arcs and it is then only checked once. The syntax check is also incremental. When the user modifies a part of a CPN diagram, the CPN editor only rechecks those objects that have been changed – or are related to changed objects. As an example, the change of a colour set means that the initialization expression and all the surrounding arc expressions have to be rechecked. If the place belongs to a fusion set or is an assigned port or socket node, it must also be rechecked that the colour sets are identical and all initialization expressions equivalent.

If the user changes the content of the declaration node, all colour sets are redeclared. This means that the entire CPN diagram has to be rechecked. To avoid using too much turn-around time for such total rechecks, the CPN editor allows the use of a temporary declaration node – in which new declarations can be added to the original declarations without enforcing a total recheck.

Use of names

Each page has a name and a number. In addition, it is possible (and advisable) to give names to each place and each transition.

Names do not influence the semantics of CP-nets. However, they are used in the feedback information from the CPN tools to the user, e.g., in the page hierarchy graph and in the hierarchy inscriptions. To make the feedback unambiguous, it is recommended to keep names unique, but this is not enforced (unless the user activates an optional syntax restriction). For places and transitions it is sufficient to have names that are unique on each individual page.

Some users tend to have a large number of transitions and places without names. This is no problem for the CPN tools, but it may make it difficult for the user to read the feedback.

Abstract data base

When the user creates a CPN diagram, the editor stores all the semantic information in an abstract data base, from which it can easily be retrieved by the CPN simulator and other analysis programs. The abstract data base was designed as a relational data base, but for efficiency it is implemented by means of a set of list structures – making the most commonly used data base operations as efficient as possible. The existence of the abstract data base makes it much easier to integrate new and existing editors and analysis programs with the CPN tools. For this purpose CPN ML provides three sets of functions. The first set reads the information of the abstract data base, e.g., the colour set of a place. The second set creates pages and auxiliary objects (which have a graphical representation, but no representation in the abstract data base). Finally, the third set converts auxiliary objects to CPN objects (which means that they become included in the abstract data base). The three sets of functions make it easy to write programs which translate CPN diagrams into textual or graphical representations of other Petri net tools, and vice versa.

6.2 Simulation of CP-nets

Simulation of CP-nets can be supported by a computer tool or it can be totally manual, for example, performed on a blackboard or in the head of a modeller. Simulation is similar to the debugging of a program, in the sense that it can reveal errors, but in practice it can never be sufficient to prove the correctness of a system. Some people argue that this makes simulation uninteresting and that the user should instead concentrate on the more formal analysis methods. We do not agree with this conclusion. On the contrary, we consider simulation to be just as important and necessary as the formal analysis methods.

In our opinion, all users of CP-nets (and other kinds of Petri nets) are forced to make simulations – because it is impossible to construct a CP-net without thinking about the possible effects of the individual transitions. Thus the proper question is not whether the modeller should make simulations or not, but whether he wants computer support for the simulation activity. With this rephrasing the answer becomes trivial. Of course, we want computer support. This means that the simulations can be done much faster and with no errors. Moreover, it means that the modeller can use all his mental capabilities to interpret the simulation results – instead of using most of his efforts to calculate the possible occurrence sequences. Simulation is often used in the design phases and the early investigation of a system design, while the more formal analysis methods are used for validation.

Integrated tool

The CPN editor and CPN simulator are two different parts of the same program and they are closely integrated with each other. In the editor it is possible to prepare a simulation, e.g., set the many options which in detail determine how the simulation is performed. In the simulator it is possible to perform simple editing operations, e.g., modify arc expressions, guards, initialization expressions, code segments and time regions. Furthermore it is possible to change the attributes of CPN objects and add auxiliary objects.

Most modellers use simulation to test the different parts of their CPN diagram as these are constructed. This is similar to a programmer who debugs the individual parts of his program as he writes them. To support this work style, the shift between the editor and the simulator must be reasonably fast. It must also be possible to simulate selected parts of a large model.

Page instances

All the instances of a page have the same net structure and the same net inscriptions. However, during a simulation the page instances will have different current markings and different sets of enabled transitions. Hence the CPN simulator must show the individual page instances, and not just the individual pages.

There is a window for each page. In this window the simulator displays all the instances of the page – one at a time. The title bar (of each page window) specifies the identity of the currently displayed page instance. The identification consists of a non-negative instance number, the page name, and a list containing all supernodes (cf. Def. 3.2 and the explanation of page instances in Sect. 3.1). The user may ask the system to display other page instances. However, during a simulation this is usually not necessary, because the CPN simulator automatically displays those instances on which transitions are occurring.

Manual and automatic simulations

To perform a simulation it is necessary to choose the binding elements to be executed. This choice can be made either by the user or by the CPN simulator. However, it is important to notice that in both cases it is the CPN simulator that performs the really complex work: the calculation of the enablings and the calculation of the effects of the occurring steps. A set of binding elements (selected for execution) is called an **occurrence set.**

In a **manual** simulation the steps are chosen by the user. However, this is done under strict supervision and guidance by the CPN simulator, which, e.g., checks the legality of all the proposed binding elements. In an **automatic** simulation the steps are chosen by the CPN simulator. This is done by means of a random number generator.

Manual construction of occurrence sets

The CPN simulator indicates the current enabling by highlighting those transition instances which have at least one enabled binding element. The highlighting may be done by increasing the line thickness of the transition, as indicated by transition T2 in Fig. 1.9. The highlighting may, however, also be done in several other ways, determined by the user (via the Feedback Options command).

To construct a step the user chooses those binding elements which he wants to include in the step. To include a binding element the user must specify a transition instance and a binding. First he selects a transition instance. This is done by a mouse click on the transition – at the correct page instance. Then the Bind command is invoked. This displays a dialogue box like the one in Fig. 6.7, where

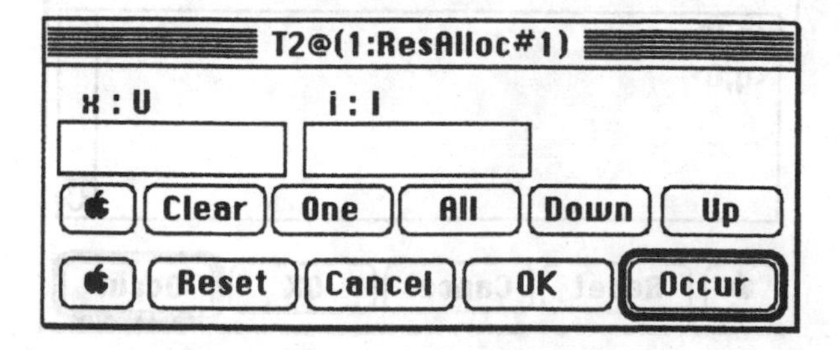

Fig. 6.7. Dialogue box for the Bind command

we assume transition T2 of Fig. 1.9 to be selected. The dialogue box has a text box for each variable of the transition. In this case there are two such variables: x of type U and i of type I. Each variable must be bound to a colour value, and this can be done in several different ways:

- One possibility is to type the desired colour values into the two text boxes by means of the keyboard. The system will check that the values are of the correct type.
- A second possibility is to press the *One* button. This will instruct the simulator to make a random choice between the enabled bindings (of the selected transition instance). The values of the chosen binding will be shown in the two text boxes.
- A third possibility is to press the *All* button. This will instruct the simulator to display all the enabled bindings (of the transition instance). This will be done in an enlarged dialogue box, as shown in Fig. 6.8. The enlarged dialogue box contains three parts. The upper part is a **working area**. It contains the same two text boxes as we had in Fig. 6.7. The middle part is a list box. It contains the **included bindings**, i.e., those binding elements (of the transition instance) which we have already included in the step. In our case, we have not yet included any such bindings, and hence the middle part of Fig. 6.8 is empty. The bottom part of the dialogue box is also a list box. It contains the **proposed bindings**, i.e., those bindings found by pressing the *All* button.

The user can move a binding from one part of the extended dialogue box to another. To do this the user selects the corresponding binding (to select the binding in the working area the user selects one of the text boxes). Then either the *Up* or *Down* button is pressed, and this will move the binding in the desired direction.

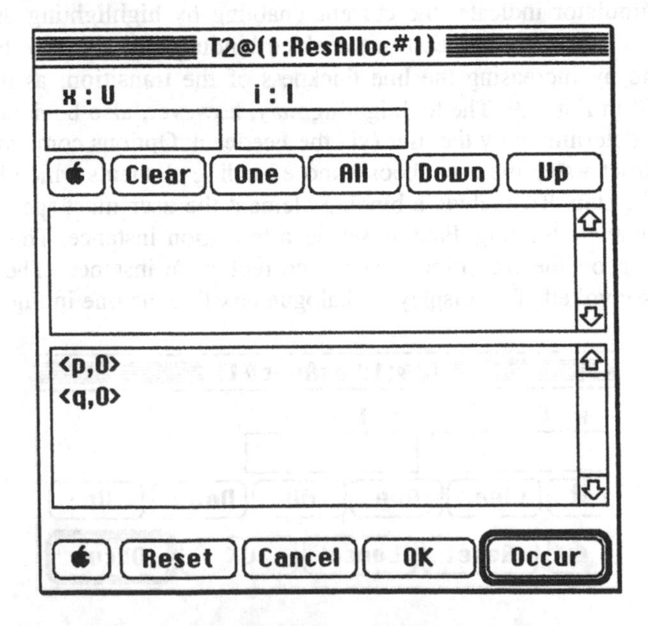

Fig. 6.8. Extended dialogue box for the Bind command

Down moves from the upper part to the middle part, from the middle to the lower, and from the lower to the upper. Analogously, *Up* moves from the lower part to the middle part, from the middle to the upper, and from the upper to the lower part.

Each time a binding element is included in the step, the simulator reserves the input tokens needed by the binding element. This is done by subtracting the tokens from the current marking. If the reservation cannot be made, the binding element cannot be included. This means that it is impossible to construct a step which is disabled in the current marking.

The CPN simulator also shows the effect of the included binding elements. This is done by displaying the set of input tokens and the set of output tokens – in key/popup regions of the corresponding arcs. Figure 6.9 shows how it looks, when in Fig. 6.8 we include the binding element (T2,<x=p,i=0>). The left-hand side shows the transition before the inclusion, while the right-hand side shows it after the inclusion (but before the occurrence). It should be noticed that the CPN simulator also recalculates the enabling. In the right-hand side of Fig. 6.9 T2 is no longer highlighted. This indicates that we cannot add more binding elements for T2 (unless we delete the one which we have already added). If there had been two additional e-tokens on S, transition T2 would have remained highlighted – and we could either include an appearance of (T2,<x=q,i=0>) or an additional appearance of (T2,<x=p,i=0>).

When the user has specified the desired binding elements for the chosen transition instance, he either presses the *OK* or the *Occur* button. When *OK* is pressed the simulator remains in the step construction phase. This means that the user can add binding elements for other transition instances. When *Occur* is pressed the constructed step is executed. The new marking and the new enabling are calculated, and after that the user starts the construction of the next step.

It is possible to mix the three ways to specify bindings. As an example, the user may type colour values in some of the text boxes (of the upper part of the Bind dialogue box) before *One* or *All* is pressed. Then the system only calculates bindings which are consistent with the typed colour values.

The manual construction of binding elements may seem to be cumbersome. However, it should be understood that manual simulations are primarily used during the design of the CPN model – in the test and validation phases most users

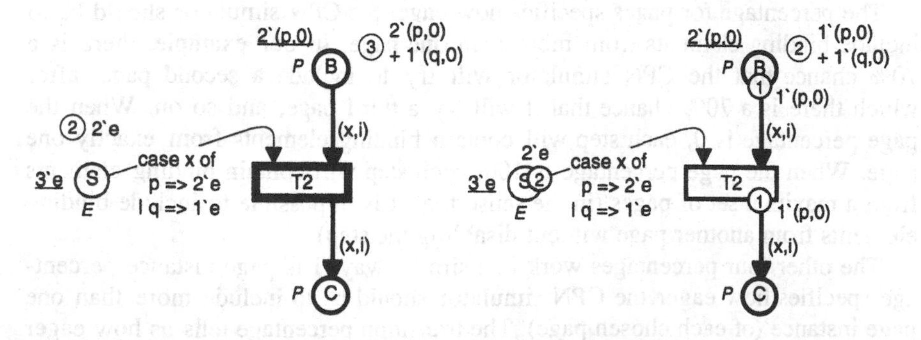

Fig. 6.9. Input and output tokens

switch to automatic simulations. Moreover, there are many shortcuts. As an example, a single key combination makes it possible to include a random binding (for the selected transition instance) without opening a dialogue box. This is useful when a transition instance has only one enabled binding – or when the user does not care which binding is chosen.

During a manual simulation the Bind command is used again and again. Thus it has been important for us to make the command as powerful as possible. This means that the dialogue box is complex and that it takes quite a while to learn all the ways in which the command can be used. The lower button makes it possible to open and close the two list boxes in Fig. 6.8. The upper button is for transitions which have so many variables that they cannot have separate text boxes and transitions which use colour sets so complex that the values do not fit the text boxes.

Automatic construction of occurrence sets

In an automatic simulation, the simulator chooses among the enabled binding elements. This is done by means of a random number generator. The selection between pages is always fair, in the sense that all pages (with enabled binding elements) have an equal chance to contribute to the step. Analogously, there is always a fair selection among the page instances (of a chosen page) and among the transitions (of a chosen page instance). For the enabled bindings (of a chosen transition) the selection may be fair or non-fair (this is determined by an option). In a fair selection the CPN simulator calculates all the enabled bindings, and then makes a fair choice between them. In a non-fair selection the CPN simulator uses the first binding it finds.

The user also specifies how large he wants the individual steps to be. As one extreme he may want each occurring step to have exactly one binding element. As another extreme each occurring step may be demanded to be maximal (in the sense that no other binding element can be added without disabling the step). The last possibility is sometimes referred to as the maximal occurrence rule, and it has been used in different kinds of Petri net models, for example, some of those which have a time concept. Between the two extremes there are many other possibilities. The detailed selection strategy is specified by the Occurrence Set Options command. The dialogue box of this command is shown in Fig. 6.10.

The percentage for pages specifies how eager the CPN simulator should be to include binding elements from more than one page. In our example, there is a 70% chance that the CPN simulator will try to include a second page, after which there is a 70% chance that it will try a third page, and so on. When the page percentage is 0, each step will contain binding elements from exactly one page. When the page percentage is 100, each step will contain binding elements from a maximal set of pages (in the sense that it is impossible to include binding elements from another page without disabling the step).

The other four percentages work in a similar way. The page instance percentage specifies how eager the CPN simulator should be to include more than one page instance (of each chosen page). The transition percentage tells us how eager the simulator should be to include more than one transition (of each chosen page instance). Finally, the two binding percentages indicate how eager the simulator should be to include different bindings and to include the same binding more than once.

When all five percentages are 100, we have the maximal occurrence rule. When they are all 0, each step contains exactly one binding element. In all other cases random drawings determine the actual size of the step, and this means that the percentages in Fig. 6.10 can both lead to a maximal step and to a step with only one binding element – although, of course, most steps will be somewhere in between. It should also be noted that the percentages influence each other. As an example, let us assume that all five percentages have a high value. This implies that each included page contributes with many binding elements and hence use many of the available tokens. Therefore it may be impossible to include more than one page (even though we have a high page percentage).

The dialogue box also allows the user to specify a seed value for the random number generator (or obtain a seed from the system clock). To repeat an automatic simulation with the same set of steps it is necessary to start with the same seed value. Otherwise, the simulator generates a different sequence of random numbers, and this means that the occurrence sequence will differ. An explanation of the remaining parts of the dialogue box can be found in the reference manual.

Observation of simulations

For a big model, it is obvious that we cannot arrange the page windows in such a way that they do not overlap. Moreover, it should be remembered that each win-

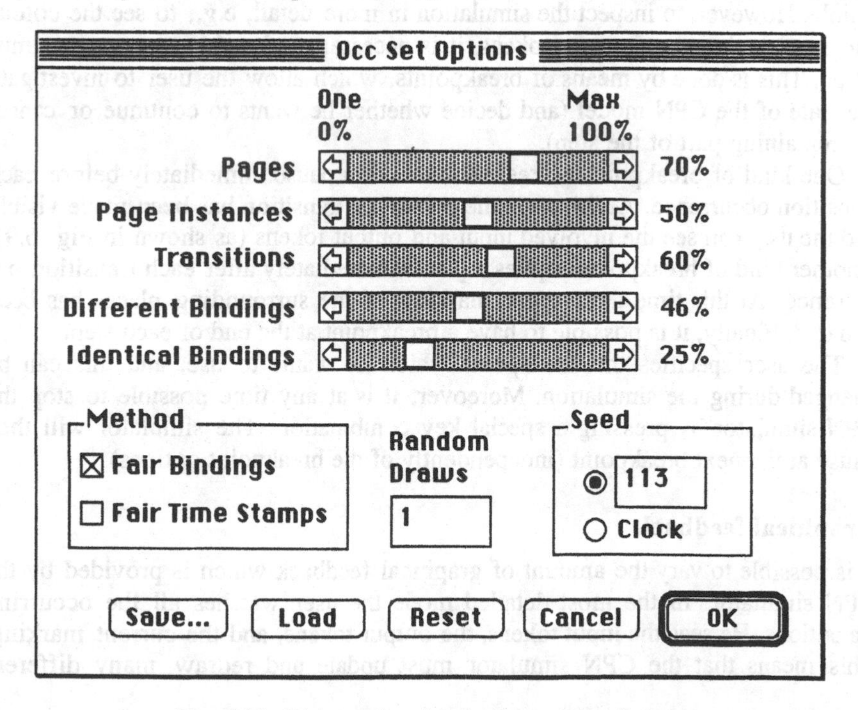

Fig. 6.10. Dialogue box for Occurrence Set Options

dow only shows one page instance at a time. Hence it is necessary that the CPN simulator – during a simulation – rearranges the screen image, so that the user can always see the interesting parts of the CPN diagram. Otherwise, it would be impossible to follow an automatic simulation.

A step may involve several transitions. Then the effects are shown for one transition at a time, and before a transition occurs, the CPN simulator automatically makes the following rearrangements:

- The interesting page is made active (and hence displayed on top of all the other windows).
- The interesting page instance is displayed in the window.
- The window is scrolled, to guarantee that the interesting transition is in the visible part of the window (this only happens if the auto-scroll option is set).

By means of one of the mode attributes the user can say that he does not want to observe all the page instances. In that case, the simulator still executes the transitions of the non-observed page instances, but without rearranging the screen. This means that the user cannot see the transitions of the non-observed page instances (unless they happen to be visible without any rearrangements). By closing a window and making all its page instances non-observable, a lot of graphical updating is avoided, and hence the simulation speed is improved.

Breakpoints

The automatic rearrangements guarantee that the occurring binding elements are visible. However, to inspect the simulation in more detail, e.g., to see the colour values of the input and output tokens, it is necessary to be able to stop the simulation. This is done by means of breakpoints, which allow the user to investigate the state of the CPN model (and decide whether he wants to continue or cancel the remaining part of the step).

One kind of breakpoint makes the simulation pause immediately before each transition occurrence. At this time the occurring transition has been made visible and the user can see the involved input and output tokens (as shown in Fig. 6.9). Another kind of breakpoint implies a pause immediately after each transition occurrence. At this time the current marking of the surrounding places has been updated. Finally, it is possible to have a breakpoint at the end of each step.

The user specifies the breakpoint which he wants to use, and this can be changed during the simulation. Moreover, it is at any time possible to stop the CPN simulator by pressing a special key combination. The simulator will then pause at the next breakpoint (independently of the breakpoint options).

Graphical feedback

It is possible to vary the amount of graphical feedback which is provided by the CPN simulator. In the most detailed mode the user watches all the occurring transitions. He sees the input tokens, the output tokens, and the current marking. This means that the CPN simulator must update and redraw many different graphical objects. This takes time and hence there is a trade-off between information and speed.

Each of the feedback mechanisms is controlled by one or more options, and thus they can be fully or partially omitted. As an example, it is possible to omit the input tokens and output tokens altogether. It is also possible to include them, but with missing popups. The user can then only see the number of input and output tokens. However, he can always stop the simulation and then double-click one of the input/output key regions. This will make the popup visible.

If the user carefully chooses those page instances which he wants to observe and those feedback mechanisms which he wants to be in effect, he can speed up the simulation – often by as much as 100%.

Selection of subdiagrams

It is important to be able to simulate selected parts of a large CPN diagram, e.g., a set of new pages, which the user has just constructed and now wants to test. Without such selection possibilities it would be difficult to build large CPN models. Simulation of a set of pages can be done by copying the pages to a new diagram. However, this is cumbersome and it easily leads to inconsistencies (because the user changes the copies without remembering to change the originals or vice versa). Hence it is important that the user can select parts of a diagram for simulation – without having to modify the diagram itself.

Some parts of a diagram can be selected by means of the multi-set of prime pages. As an example, consider the ring protocol from Sect. 3.1. If we use *1`Site#11* as the multi-set of prime pages we get one instance of *Site#11* and no instances of *Network#10*. This means that we test *Site#11* without considering *Network#10*.

However, we may also want to do the opposite, i.e., test *Network#10* without considering *Site#11*. This turns out to be slightly more difficult, and it cannot be done by the prime pages alone – because for each instance of *Network#10* we always get four instances of *Site#11*. Instead, we need a mechanism which allows the user to include a page without including its subpages. This is achieved by means of one of the mode attributes. It allows the user to specify whether a given page instance is included in the simulation or not. If it is not included, the simulation is performed as if the page instance did not exist. This means that the direct supernode becomes an ordinary transition, where the arc expressions and guard determine the enabling and the occurrence.

In the data base example, we can use *1`Network#10* as the multi-set of prime pages and we can specify that none of the subpage instances corresponding to the supernodes *Site1*, *Site 2*, *Site3* and *Site4* shall be included. Then we will have one instance of *Network#10* and no instances of *Site#11*, and hence we can test *Network#10* without considering *Site#11*. This will only be useful if arc expressions are added to *Network#10*.

Code segments

When we simulate a CP-net, it is sometimes convenient to be able to equip some of the transitions with a **code segment**, i.e., a sequential piece of code which is executed each time a binding element of the transition occurs.

As an example, consider Fig. 6.11 which shows a code segment for the transition *Free@EstabCon#2* in Fig. 3.6. The small C box is a key region, while the dashed box is the popup region. The first line of the popup contains an **input pattern**. It shows that the code segment is allowed to use (but not change) the values of the variables x and y. The remaining four lines contain the **code action**.

The code segment of Fig. 6.11 is used to maintain a graphical representation of the system state, like the one shown in Fig. 6.12. There is a rounded box for each of the telephones, i.e., for each colour in U. The text of the rounded box describes the current state of the phone. The border indicates whether the phone is inactive (thin line), used as a sender (thick full line) or used as a recipient (thick shaded line). Finally, there are three different kinds of connectors. They indicate *Requests* (thin dashed line), *Calls* (thin full line) and *Connections* (thick line).

When *Free* occurs, the x-token moves to *Long*, the y-token to *Ringing* and the (x,y)-token to *Call*. The corresponding changes of the graphical representation is performed by the code action – by means of three function calls. The remaining transitions have similar code segments. We have added an extra transition *Initialize* which occurs once (at the start of the simulation). The code segment of *Initialize* creates the rounded boxes. Finally, we have specified that we do not want any graphical feedback on the CPN diagram itself. Instead we follow the simulation by watching the diagram in Fig. 6.12 – from which we can deduce the present state of each phone and the activities in which it takes part.

For practical applications, code segments have turned out to be extremely useful. They are primarily used for three different purposes:

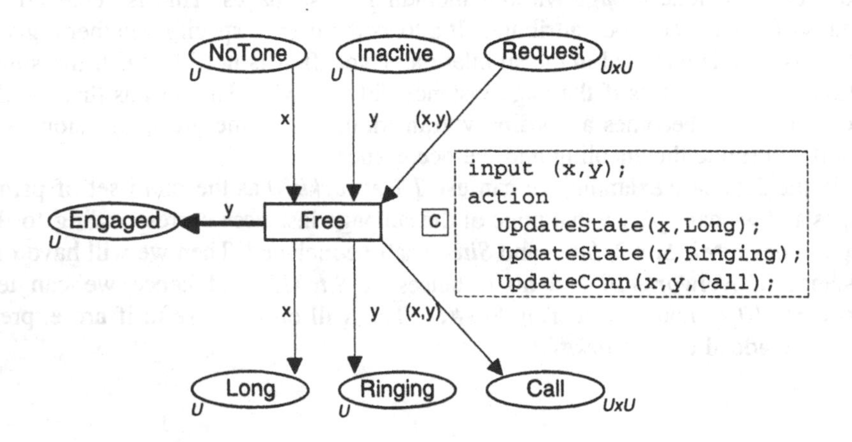

Fig. 6.11. Code segment

- Data can be read from an input file. This makes it possible to have a simulation model which can be run on different sets of data without changing the initialization expressions. It is also possible to display a dialogue box, in which the user can type the data.
- Data can be written on an output file. This makes it possible to save data, e.g., for later analysis in a spreadsheet program.
- Graphical representations of the modelled system can be made. This can be done by using ad hoc SML code (as in Figs. 6.11 and 6.12) or by using a set of predefined reporting facilities (to be introduced later in this section).

Above we have described three different uses of code segments. None of them influences the flow of tokens, and hence they do not change the set of possible occurrence sequences or the set of reachable markings. However, it is also possible to use code segments in a way where this is no longer true:

- A variable of the transition can be determined by the code segment – instead of being bound in the ordinary way. To do this the code segment must have an **output pattern**, which contains the variables to be bound by the code segment. The output pattern has a similar format as the input pattern, and can only contain variables which do not influence the enabling. Hence it cannot contain variables which appear in the input arc expressions or in the guard.
- Code segments may use reference variables, which can be read and updated. This possibility should be used with great care. It implies that two transitions can influence each other – even when they are totally unrelated (with respect to the net structure).
- A code segment may contain a special **code guard** which replaces the ordinary guard, whenever the code segment is in effect.

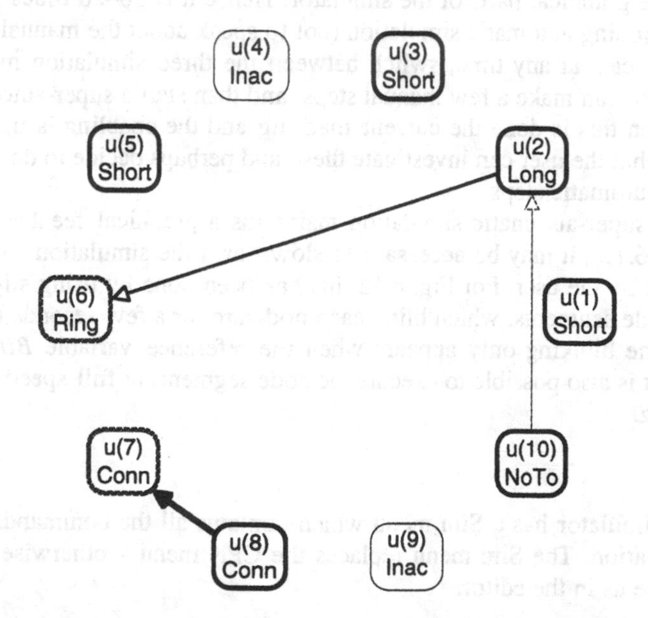

Fig. 6.12. Graphical representation maintained via code segments

By means of an option it is possible to tell the CPN simulator whether it should simulate with code segments or not. Moreover, one of the mode attributes makes it possible to execute each page instance with or without its code segments. The input and output patterns provide a well-defined interface between the code segments and the variables of the CP-net. Hence there is a well-defined and easy-to-understand relationship between a simulation with code segments and the corresponding simulation without.

In the standard version of the CPN simulator, the codes segments have to be written in the SML language. However, there is also an experimental version in which the code segments are written either in SML or in an object oriented language called Beta. It is possible to create Beta objects via code segments, and references to Beta objects can be passed via token colours.

In later versions of the CPN simulator, it will be possible to write the code segments in other programming languages, e.g., C++, Pascal, Prolog or a mixture of these. In the Unix and X-windows implementation, it is already possible to have SML code segments which use object code produced by other compilers.

Super-automatic simulations

In one of the previous subsections we saw that code segments can be used to create feedback – dumped on a file or displayed in a separate window (such as Fig. 6.12). These possibilities mean that it may be unnecessary to look at the CPN diagram itself – during an automatic simulation. Why then keep the diagram updated? This line of thought has led to the creation of a simulation mode, called **super-automatic** simulation, performed by a stand-alone SML function. The function is created by the CPN simulator, but it is executed without any interaction with the graphical parts of the simulator. Hence it is 10–50 times faster than the corresponding automatic simulation (not to speak about the manual one).

The user can, at any time, switch between the three simulation modes. This means that he can make a few manual steps, and then start a super-automatic execution. When this is done the current marking and the enabling is updated, and this means that the user can investigate these and perhaps decide to do some more manual or automatic steps.

When a super-automatic simulation maintains a graphical feedback (like the one in Fig. 6.12), it may be necessary to slow down the simulation so that it can be followed by the user. For Fig. 6.12 this has been done by using slightly more complex code segments, which blink each node/arc for a few seconds before it is changed. The blinking only appears when the reference variable *Blink* is true, and hence it is also possible to execute the code segments at full speed, i.e., without blinking.

Sim menu

The CPN simulator has a Sim menu which contains all the commands used during a simulation. The Sim menu replaces the CPN menu – otherwise the menus are the same as in the editor.

We have already described the Bind command. It allows the user to construct occurrence sets, to be used during manual simulation. However, it is not always necessary to start such a construction from scratch. Instead the user can invoke a command that proposes an occurrence set, calculated in the same way as for an automatic simulation. The proposed occurrence set can be investigated, and perhaps modified, before it is executed.

There are of course also commands to start, stop, continue and cancel simulations. A stop is a temporary pause – during which the user can inspect the current marking and the input/output tokens. When this has been done, the user either continues the simulation or cancels the remaining parts of it. An automatic simulation runs until it is stopped by the user, or until a marking with no enabled binding elements is reached. In a later version, it will also be possible for the user to specify different kinds of **stop criteria**. When one of these becomes satisfied, the simulation is stopped and the user is informed about the reason.

The Change Marking command makes it possible to add and remove tokens during a simulation. This is very convenient, e.g., when some of the initialization expressions have been forgotten. The command can also be used to make a detailed test of a transition. The modeller can put different tokens on the input places and then use the *All* button of the Bind command to test that the transition has the expected enabling. There is also a command which allows the user to return to the initial marking, i.e., forget all the steps which have been executed.

Finally, there are commands to generate and start super-automatic simulations and commands to save and load system states via files. The latter make it possible to save an interesting marking, and then later return to it without having to repeat the steps.

Simulations with time

To investigate the **performance** of systems, i.e., the speed by which they operate, it is convenient to extend CP-nets with a time concept. To do this, we introduce a **global clock**. The clock values represent the **model time**, and they may either be discrete (which means that all clock values are integers) or continuous (which means that the clock values are reals). In addition to the token colour, we allow each token to have a **time stamp** attached to it. The time stamp describes the *earliest* model time at which the token can be removed by a binding element.

The execution of a timed CP-net is time driven, and it works in a similar way as the event queues found in many programming languages for discrete event simulation. The system remains at a given model time as long as there are enabled binding elements which are ready for execution. Then the system advances the clock to the next model time at which enabled binding elements can be executed. Each marking exists in a closed interval of model time (which may be a point, i.e., a single moment).

To see how a time simulation works, let us consider Fig. 6.13 which contains a timed CP-net for the resource allocation system from Sect. 1.2. The first line of the declarations specifies that we work with a discrete clock, starting at 0. From the fourth and fifth lines, we see that P-tokens are timed (i.e., carry time stamps), while the E-tokens do not. This means that E-tokens are always ready to be used. The small rectangle below the declarations indicates that the current model time is 641 (in the CPN simulator this information is displayed in the status bar). The marking of place A contains two tokens, one with colour (q,4) and time stamp 627, and one with colour (q,5) and time stamp 598. Analogously, place B has two tokens, one with colour (p,6) and time stamp 567, and one with colour (q,1) and time stamp 602. Place D has a single token with colour (p,4) and time stamp 641. Finally, place T has one token with colour e and no time stamp. The @ signs in the current markings should be read as "at". Each @ sign is followed by a list of time stamps. In Fig. 6.13 all lists have length 1, because all the tokens have different colours. However, the initial marking of place A is displayed as 3\`(q,0)@[0,0,0], while the initial marking of B is 2\`(p,0)@[0,0].

Now let us consider the steps which may occur in the markings of Fig. 6.13. The binding element b_1 = (T4,<x=p,i=4>) is enabled, because it satisfies the requirements of the usual enabling rule (Defs. 2.8 and 3.6). The binding element is also **ready**, because all the time stamps of the involved input tokens are smaller than or equal to the current model time (in this case there is only one such time

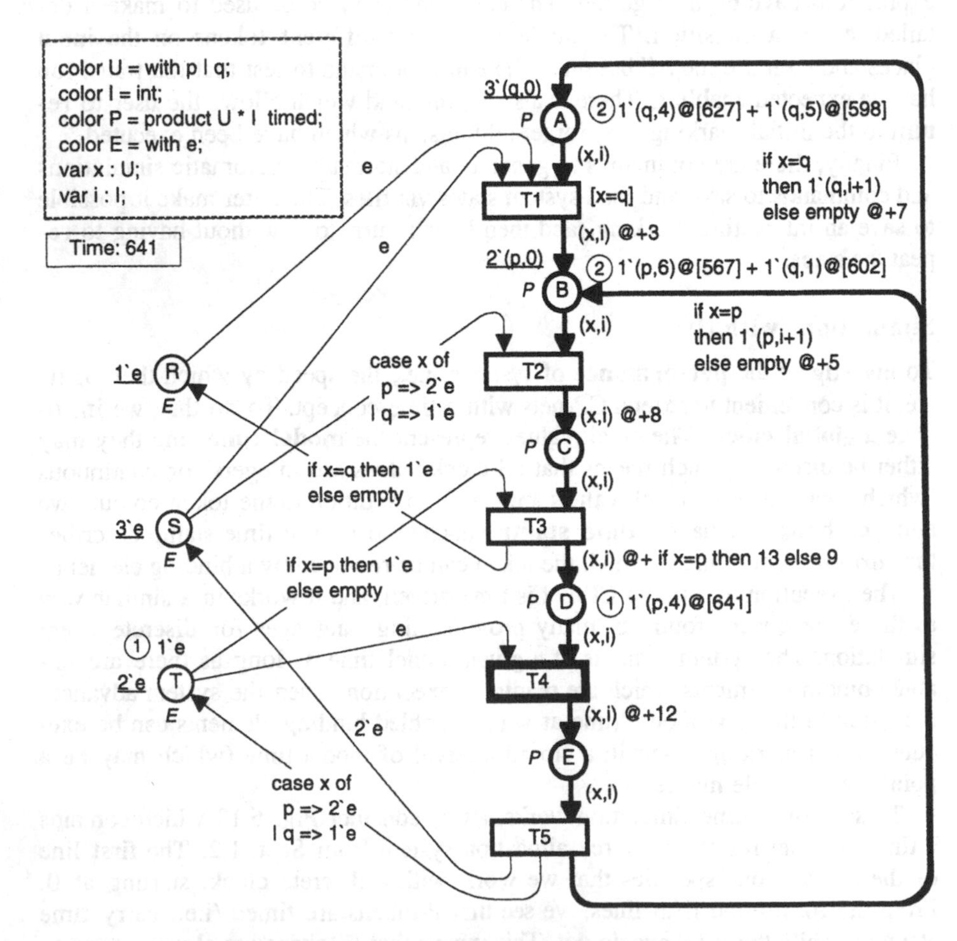

Fig. 6.13. Timed CP-net for the resource allocation system

stamp and it is equal to the current model time). When a binding element is enabled and ready it may occur. The occurrence of b_1 removes the token from D and the token from T. The occurrence of b_1 will also add a token to E. The colour of the new token is calculated in the usual way, i.e., as specified by the occurrence rule (Defs. 2.9 and 3.6). The time stamp of the new token is calculated as the current model time plus a **time delay**, which is specified in the **time clause** of the corresponding output arc expression – after the @+ operator. In this case the delay is 12 time units, and hence we get 653 as the new time stamp. Intuitively, this means that the p-token must stay at least 12 time units at place E. We can interpret this to mean that the state E has a minimal duration of 12 time units. We can also interpret it to mean that the activity T4 takes 12 time units.

When b_1 has occurred, we reach a marking in which b_2 = (T5,<x=p,i=4>) is the only enabled binding element. However, b_2 is not ready at model time 641 because it involves a token with a too-high time stamp. Hence we increase the model time, until b_2 becomes ready, i.e., to 653. If there had been several en-

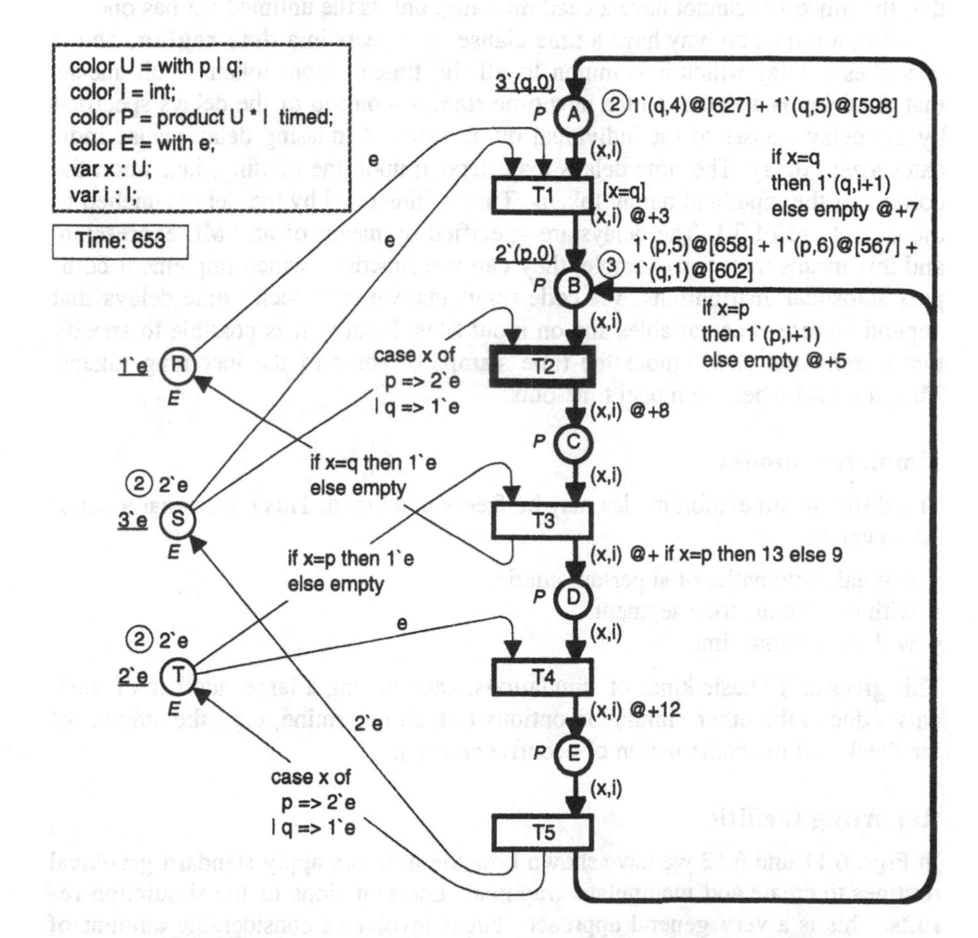

Fig. 6.14. A second marking of the timed resource allocation system

abled binding elements we would have increased the model time until one of them became ready. When b_2 has occurred, at model time 653, we get the marking shown in Fig. 6.14. Now we have three enabled binding elements $b_3 = (T2,<x=p,i=5>)$, $b_4 = (T2,<x=p,i=6>)$ and $b_5 = (T2,<x=q,i=1>)$. The binding elements b_4 and b_5 are ready at 653, while b_3 uses a token with time stamp 658. Hence either b_4 or b_5 will occur. They are in conflict with each other, because they both need e-tokens from S. This means that only one of them will be executed. However, had there been an additional e-token on S, b_4 and b_5 would have been concurrently enabled and both of them would have occurred at time 653 (either in the same step, or in two subsequent steps).

For a timed CP-net we demand that each step consists of binding elements which are both enabled and ready. Hence the possible occurrence sequences of a timed CP-net always form a subset of the possible occurrence sequences of the corresponding untimed CP-net. This means that we have a well-defined and easy-to-understand relationship between the behaviour of a timed CP-net, and the behaviour of the corresponding untimed CP-net. As an example, it is easy to see that the timed net cannot have a dead marking, unless the untimed net has one.

Also, a transition may have a time clause. It appears in a **time region**, and it describes a delay which is common to all the timed output tokens. This means that the delay is added to all the new time stamps – on top of the delays specified by the delay clauses of the individual output arcs. A missing delay clause indicates a zero delay. The time delays may depend upon the binding, i.e., upon the colours of the input and output tokens. This is illustrated by the delay attached to the output arc of T3. The delays are specified by means of an SML expression, and this means that, for example, they can use functions which implement complex statistical distributions. Via code segments, we can specify time delays that depend on reference variables and on input files. Finally, it is possible to specify that a transition shall ignore the time stamps of some of the incoming tokens. This is useful when we model time-outs.

Simulation modes

The different simulation modes may be freely combined. This means that a simulation can be:

- manual, automatic, or super-automatic,
- with or without code segments,
- with or without time.

This gives us 12 basic kinds of simulations, each having a large number of variants – due to the other simulation options (which determine, e.g., the amount of feedback and the construction of occurrence sets).

Reporting facilities

In Figs. 6.11 and 6.12 we have shown how the user can apply standard graphical routines to create and manipulate graphical representations of the simulation results. This is a very general approach, but it involves a considerable amount of work. The modeller has to write code segments which create and manipulate the graphical objects – via a set of high-level graphical functions.

However, the CPN simulator also contains a set of much more straightforward reporting facilities providing the usual kinds of business graphics, e.g., bar charts, pie charts and function charts. To create a chart the user invokes the Chart command. He specifies the desired kind of chart and then gets a dialogue box in which he can specify all the details of the chart. For a bar chart it is necessary, e.g., to specify the number of bars, indicate whether the corresponding values are integers or reals, and specify how the axes and grid lines should look.

The user also writes a code segment for the chart. The code segment is executed each time the chart is refreshed, and it calculates the individual data values to be displayed in the chart. The data values are calculated by means of SML expressions, and they can, e.g., use the marking of a given place (via a set of predefined functions). When the desired data values have been calculated they are inserted into the **chart data structure**, which was created by the Chart command (together with the chart). The data structure is either a vector or a matrix, and from this data structure the system automatically calculates the actual appearance of the chart. The code segment also specifies the refresh frequency. As an example, the user may say that he wants the chart to be updated for each 100 steps, or for each 1000 time units (of a timed simulation).

For the resource allocation system we can create the bar chart shown in Fig. 6.15. It describes the delay of p-processes and q-processes due to the sharing of resources. The values of Fig. 6.15 correspond to the system state shown in Fig. 6.13. In this state each of the three q-processes has existed for 641 time units. If there had been enough resources they would have been able to make a full cycle for each 39 time units. This is found by adding the delays of the five transitions. However, due to the lack of resources, the q-processes have made many fewer cycles. Together they have made 10 cycles plus an additional step from A to B. This corresponds to 393 time units, out of the $3*641$ which were available.

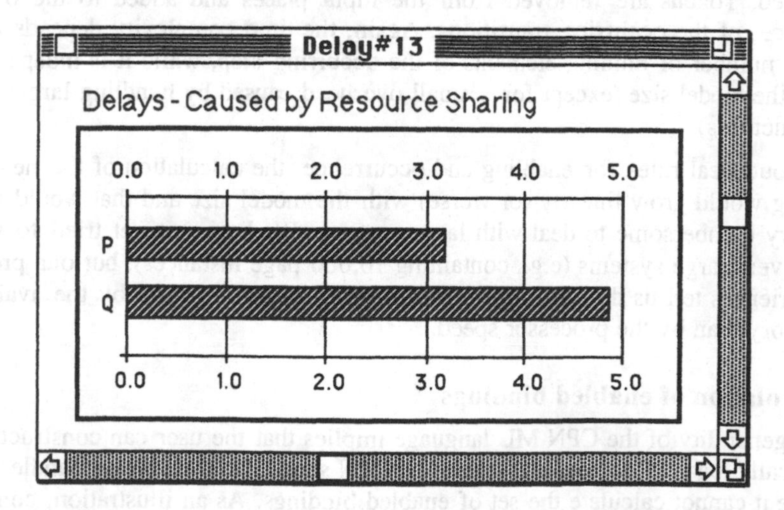

Fig. 6.15. Example of the more elaborated reporting facilities

Hence we conclude that the lack of resources has caused a delay, which for the q-processes reduces the speed by a factor: $3*641/393 = 4.89$. A similar argument gives us that the speed of p-processes is reduced by: $2*641/401 = 3.20$.

The creation of the bar chart and the redrawing of it is automatic. The user only has to write the functions which calculate the delays.

Simulation of large CPN diagrams

The CPN simulator is designed to work with complex CP-nets. Some of the industrial applications described in Chap. 7 use CPN models which have 150 page instances, each with 5–25 CPN nodes. Fortunately, it turns out that a CP-net with 100 page instances usually simulates almost as fast as a CP-net with only a single page instance (when speed is measured in terms of the number of occurring binding elements). To understand why this is the case, let us consider the three phases performed by the CPN simulator during the execution of a step:

- In the first phase, the new enabling is calculated. The enabling and occurrence rules of CP-nets are strictly local – in the sense that the rules for each transition only involve the surrounding places (plus the places related to the surrounding places by a fusion set or a port assignment). This means that the new enabling is identical to the previous enabling except for those transitions which are in the nearest neighbourhood of the transitions in the occurring step. Hence the time complexity of the enabling calculation depends upon the number of binding elements of the occurred step, while it is independent of the model size (except for a small overhead caused by handling larger data structures).
- In the second phase, the step is constructed (as described earlier in this section). The construction is much faster than the enabling calculation and the execution of the step. Hence we can ignore the time used for it.
- In the third phase, the step is executed, and this means that the marking is updated. Tokens are removed from the input places and added to the output places of the occurring transitions. Again, the time complexity depends upon the number of binding elements of the occurring step, while it is independent of the model size (except for a small overhead caused by handling larger data structures).

Without local rules for enabling and occurrence, the calculation of the new enabling would grow linearly (or worse) with the model size and that would make it very cumbersome to deal with large systems. We have not yet tried to work with very large systems (e.g., containing 10,000 page instances), but our present experiences tell us that the upper limit is more likely to be set by the available memory than by the processor speed.

Calculation of enabled bindings

The generality of the CPN ML language implies that the user can construct syntactically legal CPN diagrams which the CPN simulator is unable to handle – because it cannot calculate the set of enabled bindings. As an illustration, consider the upper leftmost transition in Fig. 6.16, where x is a variable of type X, while f∈[X→A] and g∈[X→B] are two functions. To calculate the set of all enabled bindings for such a transition it is either necessary to try all possible values of X or use the inverse relations of f and g. However, this may not be possible, because X may have too may values while the inverse relations may be unknown.

To avoid such problems the CPN simulator demands that each variable v of a transition fulfils at *least* one of the following conditions:

(i) v appears in an input arc expression which is simple enough to be used to bind v. This means that the arc expression must be a **pattern**, i.e.:
  - a single variable, e.g., v,
  - a tuple, e.g., (v,3) or (x,v),
  - a record, e.g., {sen=v,rec=S(3)},
  - a list, e.g., v::tail,
  - a union, e.g., Floor(v).

  It is allowed to nest tuples, records and lists inside each other, e.g., (x,(v,y))::tail, and it is also allowed to use the same variable repeatedly, e.g., (v,v) or v::(v::tail). Finally, it is allowed to combine expressions by means of the ` and + operations, e.g., 2`(v,x)+1`(v,y).

  A simple arc expression cannot contain functions or other operations. This means that f(v) cannot be used to bind v, while 1`(f(v),x)+2`(y,z) can only be used to bind y and z.

(ii) v has a colour set which is so small that it is possible to try all the values. All the bool, unit, enumeration or index colour sets are considered to be small, and so are product, record, union and subset colour sets iff they are built from small colour sets.

(iii) v appears in a guard [..., ptn = expr,...] or [..., expr = ptn,...] where ptn is a pattern (without ` and +), while expr is an expression in which all variables satisfy (i), (ii) or (iii).

(iv) v *only* appears in output arc expressions. This means that the v does not influence the enabling, and hence we can use all possible colour values.

It is very seldom that the demands in (i)–(iv) present any practical problems. Most net inscriptions fulfil the conditions, and otherwise they can usually be rewritten without changing the semantics. As an example, consider the three transitions in the upper part of Fig. 6.16 (the exp function takes two non-negative integers x and y as arguments and returns x^y). The CPN simulator cannot handle any of the three upper transitions. For T1 it is impossible to bind x, while for T2 it is z and for T3 n which causes the problems. Now let us assume that the function f has an inverse function fI∈[A→X]. Then we can rewrite the three transitions without changing the semantics. The modified transitions are shown in the lower part of Fig. 6.16. In the new T1 we have introduced a new variable v of type A. For the new T2 we have the same variables as before. However, we have made a trivial rewriting of the guard, and this means that z can now be bound by the guard (as soon as x and y have been bound by the arc from P4). For the new T3 we use a function sq which takes a non-negative integer x as argument and returns the integer which is closest to $\sqrt{x}$.

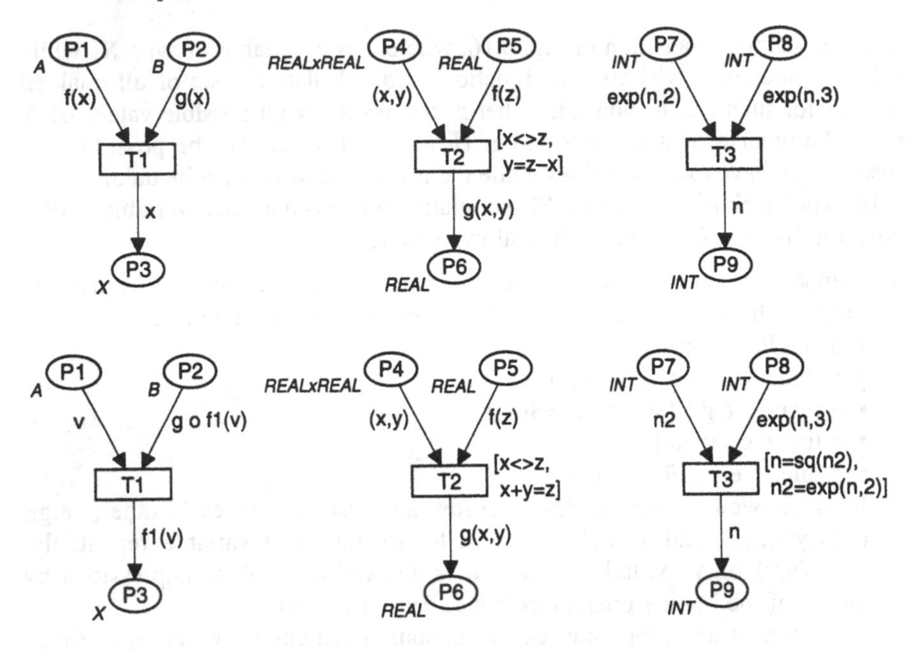

Fig. 6.16. The CPN simulator cannot handle the upper transitions, but it can handle the lower ones

Syntax for CP-net inscriptions

It is important to remember that the definition of CP-nets only talks about expressions and colour sets, without specifying a syntax for these. It is only when we want to implement tools (or present examples) that we need a concrete syntax. CPN ML has been developed for the CPN tools. However, other implementations of CP-nets may use different inscription languages – and still they deal with CP-nets.

6.3 Computer Tools for Formal Analysis

This section describes a CPN tool which supports the construction and analysis of occurrence graphs. The tool is rather new, and hence we have less experience with it than we have with the CPN editor and the CPN simulator. At the end of the section, we also give a brief description of two other CPN tools, which we intend to implement over the next years. One of these tools supports analysis by means of place and transition invariants, while the other supports the use of reduction rules.

Construction of occurrence graphs

The present version of the occurrence graph tool allows the user to construct and analyse full occurrence graphs for hierarchical CP-nets. The construction is totally automatic, and the user can tell whether he wants the entire occurrence graph to be calculated or only a part of it. This is done by means of a set of stop criteria and a set of branching criteria.

The **stop criteria** tell us how large the constructed graph should be. As an example, it is possible to specify that the construction should finish when a certain amount of CPU time has been used or when a certain number of nodes have been constructed. It is also possible to check the nodes as they are being processed, e.g., to say that the construction should finish when 3 dead markings have been found. There are many other kinds of stop criteria and this means that the user can control the construction in great detail.

The **branching criteria** imply that we do not develop all the successor markings of a given node. As an example, the user may specify that he wants the construction to investigate at most two binding elements (for each enabled transition instance). When a partial graph has been constructed, by using a set of stop criteria or a set of branching criteria, it is later possible to continue the construction (with or without stop and branching criteria).

Analysis of occurrence graphs

The occurrence graph tool is written in SML, and the constructed graph is a complex SML data structure. This means that it can be easily analysed. For this purpose we have implemented a function which is called **SearchNodes**. It takes six arguments:

- The first argument specifies the **search area**, i.e., the part of the graph which should be searched. This can, e.g., be the entire graph or a strongly connected component (see below).
- The second argument specifies the **predicate function**. It maps each node into a boolean value. Those nodes which evaluate to false are ignored; the others take part in the further analysis.
- The third argument specifies the **search limit**. It is an integer, and it tells us how many times the predicate function must be true before we terminate the search.
- The fourth argument specifies the **evaluation function**. It maps each node into a value, of some type A. The evaluation function is used to those nodes (of the search area) for which the predicate function is true.
- The fifth argument specifies the **start value**. It is a constant, of some type B.
- The sixth argument specifies the **combination function**. It maps from A×B into B, and it describes how the individual results (obtained by the evaluation function) is combined to yield the final result.

It should be noticed that the predicate function, the evaluation function and the combination function are all written by the user. This means that they can be arbitrarily complex.

SearchNodes may seem a bit complicated. However, it is also extremely general and powerful. As an example, we can use it to test whether there are dead markings. Then we use the following arguments:

- Search area: Entire occurrence graph.
- Predicate function: fun Pred(n) = (length(OutArcs(n)) = 0).
  We use a predefined function to get a list of the outgoing arcs. If the length of this list is zero there are no successors, and thus we have a dead marking.
- Search limit: This can, e.g., be 10 which means that we stop the search when 10 dead markings have been found.
- Evaluation function: fun Eval(n) = n.
  We use the identity function which maps a node into itself.
- Start value: [ ].
- Combination function: fun Comb(new,old) = new::old.
  We add the new result to the list of the previous results.

With these parameters SearchNodes returns a list with at most 10 dead markings. If the list is empty there are no dead markings. If the length is less than 10, the list contains all the dead markings.

When we deal with a partial occurrence graph in which some of the nodes have not yet been processed, it is better to have a slightly more complex predicate function. We then avoid those nodes which we have not yet processed. This is done by using a predefined function to test the node status:

$$\text{fun Pred(n)} = (\text{length(OutArcs(n))} = 0 \wedge \text{CalcStat(n)} = \text{Processed}).$$

As a second example, we may use SearchNodes to find the best possible upper multi-set bound of a given place instance p'. This is done by using the following arguments:

- Search area: Entire occurrence graph.
- Predicate function: fun Pred(n) = true.
- Search limit: ∞.
- Evaluation function: fun Eval(n) = Mark(p')(n).
  We use a predefined function to get the marking of the place instance p' at the node n.
- Start value: Ø.
- Combination function: The function which maps two multi-sets $m_1, m_2 \in S_{MS}$ into a multi-set $m \in S_{MS}$ such that $m(s) = \max(m_1(s), m_2(s))$ for all $s \in S$. This is a general multi-set operation, and the function is predefined.

It should be obvious that we can find all the other kinds of bounds in a similar way (i.e., lower bounds, integer bounds and bounds determined by a function).

In addition to SearchNodes there is a function called **SearchArcs**. It takes the same six arguments, and it works in a similar way – except that it searches arcs, instead of nodes.

Strongly connected components

Let an occurrence graph OG with nodes N be given. A **strongly connected component** of OG is a *maximal* set of nodes $c \subseteq N$, such that all elements of c are reachable from each other (in OG). The set of all strongly connected components is denoted by SCC. It can be proved that the strongly connected components are always pairwise disjoint.

For each occurrence graph OG we define a directed graph, called the **SCC graph**. It has a node for each component $c \in SCC$, and there is an arc from the component $c_1 \in SCC$ to the component $c_2 \in SCC$ iff there exists nodes $n_1 \in c_1$ and $n_2 \in c_2$ such that n_1 has an arc to n_2 (in OG). It can be proved that the SCC graphs are always acyclic. SCC nodes without incoming arcs are said to be **initial**, while those without outgoing arcs are **terminal**. It can be proved that the SCC graphs always have exactly one initial node (which contains the initial marking of OG). SCC graphs are often very small. Most CP-nets considered in this book have an occurrence graph with a single strongly connected component, and hence the SCC graph has a single node and no arcs.

The occurrence graph tool can construct the SCC graph and display it graphically. The calculation is done by means of a standard algorithm from graph theory. The display uses a simple layout algorithm (which only works well when there are relatively few nodes). Figure 6.17 shows the SCC graph for two different CP-nets, CPN_1 and CPN_2. The initial SCC node is drawn with a thick line, while the terminal SCC nodes are drawn with a double line.

The terminal nodes of an SCC graph can be used to determine whether the corresponding CP-net has home markings or not. For CPN_1 there are no home markings. The proof is by contradiction. If a home marking M exists, the corresponding node N_M must belong to c_5 – otherwise M cannot be reached from the markings of c_5. A similar argument tells us that N_M also must belong to c_6. However, this is a contradiction, because c_5 and c_6 are known to be pairwise disjoint. Hence we conclude that the assumption (of a home marking) is wrong.

Using similar arguments as above, we can prove that a CP-net has home markings iff the SCC graph has exactly one terminal node – in which case the home markings are those markings which belong to the terminal node. Hence we conclude that CPN_2 has the markings of c_4 as home markings, and no others. It can also be proved that a set of markings is a home space iff it contains at least one marking from each terminal node of the SCC graph. Hence we conclude that CPN_1 has home spaces with only two markings.

The terminal nodes of a SCC graph can also be used to determine whether a given binding element $(t',b) \in BE$ is live or not. To do this we simply have to check whether the binding element appears on an arc in *each* terminal node, and this can be done by using SearchArcs with the following arguments:

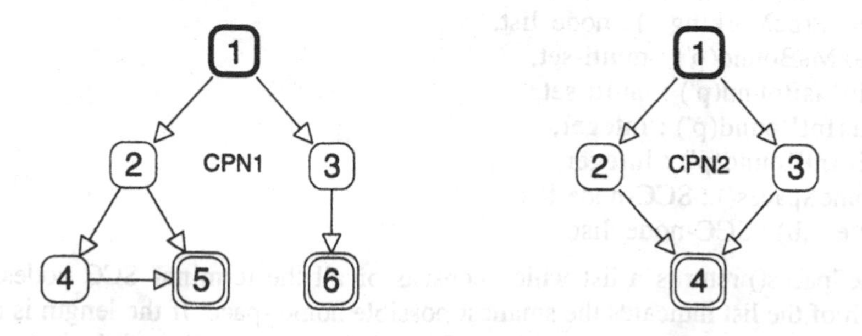

Fig. 6.17. SCC graphs for two different CP-nets

- Search area: The arcs of the terminal strongly connected component which we want to test.
- Predicate function: fun Pred(a) = (b ∈ Bind(t')(a)).
  We use a predefined function to get the bindings of the transition instance t' at the arc a.
- Search limit: 1.
- Evaluation function: fun Eval(a) = true.
  Remember that the evaluation function is only used for those arcs which satisfy the predicate function.
- Start value: false.
- Combination function: fun Comb(new,old) = new.
  This returns the value of the last evaluation.

The terminal nodes of the SCC graph can also be used to determine whether there are dead markings. For a fully constructed occurrence graph this gives us a much faster check than the one described in the previous subsection. We simply test whether there are terminal SCC nodes which only contain one occurrence graph node.

Standard queries

Above we have shown how SearchNodes, SearchArcs and the SCC graph can be used to make an automatic check of some of the dynamic properties of a CP-net. Similar tests can be made for the other properties mentioned in Sects. 4.2–4.4. For the fairness properties of Sect. 4.5 the necessary tests are more complicated, and we do not yet know to what extent these properties can be automatically determined from the occurrence graph.

For each of the examples above, we have shown the six arguments of SearchNodes and SearchArcs, using a notation which is very similar to the actual SML code used by the occurrence graph tool. Each call of SearchNodes or SearchArcs typically uses 2–5 lines of SML code, and they can be written without a deep knowledge of SML.

However, most users do not even need to write the arguments themselves. Instead they use a set of predefined SML functions, implementing the most common standard queries, e.g.:

- DeadMarkings() : node list,
- DeadProcMarkings() : node list,
- MaxMsBound(p') : multi-set,
- MinMsBound(p') : multi-set,
- MaxIntBound(p') : integer,
- MinIntBound(p') : integer,
- HomeSpaces() : SCC-node list,
- Live(t',b) : SCC-node list.

HomeSpaces() returns a list which consists of all the terminal SCC nodes. The length of the list indicates the smallest possible home space. If the length is larger than 1 there are no home markings. Live(t',b) returns a list of those terminal

SCC nodes in which (t',b) does not appear. If the list is empty (t',b) is live; otherwise the list indicates where the problems are.

Graphical display of occurrence graphs

The constructed occurrence graphs are usually very large. With the present version of the tool, we have constructed graphs which have 25,000 nodes and 100,000 arcs. With the final version, we expect to be able to handle much larger graphs. Usually, it does not make sense to display the entire graph. However, it often makes sense to display selected parts of it, e.g., the nearest surroundings of some interesting nodes/arcs found via SearchNodes or SearchArcs.

The occurrence graph tool supports the drawing of partial (or full) occurrence graphs which look similar to those shown in Figs. 5.1 and 5.2. The user specifies the nodes which he wants the tool to display. This can be done by explicitly listing the desired nodes. It can also be done by asking for the successors or predecessors of some already displayed nodes. The new nodes are drawn with a default position and with default attributes. The objects of the occurrence graph are ordinary graphical objects, and this means that the user can change the attributes. Usually, he will modify the positions as he adds new nodes – to get a nice layout without too many crossing arcs.

The user determines the text strings to be displayed for the individual nodes and arcs. This is done by specifying two SML functions. One of these maps from nodes into text strings, while the other maps from arcs into text strings. Usually, the first function gives a condensed representation of the marking (of the node), while the second gives a condensed representation of the binding elements (of the arc). Each text string is displayed in a region of the corresponding node/arc, and the region can be made visible/invisible by double clicking the node/arc. In Figs. 5.1 and 5.2 we have made all the regions visible. Each node region is positioned on top of the corresponding node, while each arc region is positioned next to the corresponding arc.

The above facilities makes it fast to draw small occurrence graphs like Figs. 5.1 and 5.2 (or small parts of a large occurrence graph). First the user writes the two text functions (determining how he wants the markings and binding elements to be displayed). This is done by applying a set of predefined SML functions. Then the graph is drawn. The occurrence graph tool makes all the calculations and it creates the necessary nodes, arcs and regions. The user only adjusts the layout and determines the extent of the drawing.

The user can split the display of a large occurrence graph into a number of different pages, e.g., one for each strongly connected component. Then the system maintains a consistent representation of the connections between the individual pages. This works in a way which is similar to the port and socket nodes used in hierarchical nets.

Integration with the CPN simulator

It is possible to make an occurrence graph for part of a large model. This is done in exactly the same way as in the CPN simulator, i.e., by means of the prime pages and one of the mode attributes. An occurrence graph can be constructed with or without time and with or without code segments. Code segments should, however, be used with care. If they have side effects which must be executed in a particular order, it makes no sense to construct an occurrence graph (because this will execute the code segments in a wrong order).

The present version of the occurrence graph tool is a separate program. However, the next versions will be more tightly integrated with the CPN simulator. This will mean that the simulator can make an automatic simulation – of an occurrence sequence which has been found in the occurrence graph tool. It will also mean that the occurrence graph tool will be able to refer to the current marking of the simulator (and, e.g., search for nodes which have an identical or a covering marking).

Other kinds of occurrence graphs

The present version of the occurrence graph tool only deals with full occurrence graphs. Later it is the intention to include the use of symmetrical markings and stubborn sets (and perhaps also covering markings). As indicated in Sect. 5.1, this will make it possible to use occurrence graphs for more complex CP-nets – because the occurrence graphs become much smaller.

We shall also implement **colour set restrictions**. The basic idea behind this concept is to be able to ignore parts of complex token colours, e.g., one or more components of a record or product colour set. As an example, it may be adequate to ignore, during the analysis of a communication protocol, the data contents of the messages – in a similar way as we ignored the cycle counters of the resource allocation system in Sect. 5.1.

Colour set restrictions are specified together with the declarations of the colour sets. This means that they can be made without changing the net inscriptions (or the net structure). Hence it is possible to have a complex CP-net, and then from this create a number of simplified models – without rewriting the net inscriptions. Colour set restrictions are not only useful for occurrence graphs. They can also be used for many other purposes, e.g., in connection with simulation and systematic debugging.

CPN tool for place and transition invariants

The first prototypes of a CPN tool for invariants analysis are under development. The final tool will be able to find and check invariants for hierarchical CP-nets. Moreover, it will assist the modeller in the use of invariants – to prove dynamic properties of the modelled system. Below we describe how to support place invariants. Transition invariants can be dealt with in a similar way.

The calculation of invariants is done in two steps. The first step is automatic and it performs a reduction of the CP-net by applying reduction rules, which are proved to preserve the invariants. There are two different sets of reduction rules. One of them preserves place invariants while the other preserves transition invariants. The second step is interactive and it is performed directly on the CPN diagram, i.e., upon the graphical representation of the hierarchical CP-net. The user proposes weights for a number of places. Typically he defines a small number of non-zero weights (for places he is interested in) and a large number of zero weights (for places that are known to be without interest for the present invariant). Then the invariant tool calculates those weights which can be uniquely determined from the weights proposed by the user. In this process the tool may also detect that some weights are inconsistent – because they violate property (**) in Sect. 5.2. Such transitions will be highlighted.

To calculate new weights and detect inconsistencies, the invariant tool uses the reduced CP-net obtained in the reduction step – but it shows the weights and the inconsistencies on the original CPN diagram. The user inspects the calculated weights and the highlighted transitions. Then he may add new weights, modify existing weights, or change the behaviour of transitions (e.g., by modifying arc expressions and guards). The process continues, with a number of iterations, and at the end an invariant will be constructed (with some weights specified by the user and the remaining ones derived by the invariant tool).

The method described above may seem primitive and cumbersome – but this is *not* the case. On the contrary, it is often possible for the user to obtain useful invariants by defining a few weights. As an example, we can find each of the four invariants in Sect. 5.2 by specifying the weights of B, R, S and T (of which at least two are zero functions). It should also be recalled that a modeller from his basic understanding of the system often has a very good idea about how the expected invariants will look. This means that he can immediately identify a lot of zero weights.

Over several years we have taught the above method to students, who have successfully used it to find invariants for small nets without having any computer support at all. The students have specified a set of proposed weights, checked the consistency and calculated derived weights. In practice, it always turned out that it was the last two things which were difficult – because they were time consuming and error-prone. With the described computer support this problem disappears because all the calculations are done by the invariant tool.

The check of a fully specified invariant is a special case of the method described above. The user simply specifies *all* the weights and the invariant tool checks the consistency.

When a set of invariants has been found, it can be used to prove system properties, and this will also be supported by the invariant tool. As an example, the user may specify the marking of some places. Then the tool can calculate upper and lower bounds for other places. This is done by using the invariants, and a reasoning which is similar to the one used in the liveness proof of Sect. 5.2. In this process the tool may determine that the specified set of place markings is inconsistent, i.e., impossible for a reachable marking. When the liveness proof is made in this way, it becomes easier, faster and more reliable – because most of the complex calculations are done by the invariant tool.

CPN tool for reduction

For non-hierarchical CP-nets there already exist different sets of reduction rules, which have been proved to be sound. However, the rules have to be modified to cope with hierarchical nets, and they have to be implemented in such a way that they can handle net inscriptions written in CPN ML. To be able to cope with complex CP-nets, it is obvious that the reductions must be performed totally automatically. This will also imply that the user does not have to know the reduction rules or the soundness proof.

To implement an automatic reduction two things are needed. First, it is necessary to choose the order in which the different reduction rules are applied, and to choose the subnets which the rules are applied to. Secondly, it is necessary to be able to check that the prerequisites of the chosen reduction rules are fulfilled, and it is necessary to be able to perform the detailed net manipulations, i.e., the replacements of subnets with other subnets. The first task is the more difficult. It involves the development of a good selection strategy. The second task is much more straightforward. It only involves the implementation of a set of calculations, which individually are quite simple.

Bibliographical Remarks

There is a large number of different groups who work with the development of Petri net tools. However, many of the tools are still research prototypes. Only a few tools handle high-level nets and very few handle hierarchical nets. For use in industrial environments, there are only a few tools that are powerful enough, sufficiently robust, and have the necessary documentation and support. A list of available Petri net tools are published at regular intervals, e.g., in *Advances in Petri Nets*. The most recent tool lists are [34] and [35]. The first contains all kinds of Petri net tools, while the second only contains tools for high-level nets.

A brief introduction to the CPN tools can be found in [1]. Information about the current versions of the CPN editor and the CPN simulator can be found in the reference manual [55]. Some of the future extensions are described in [57] and [62]. The occurrence graph tool is described in [56], the Beta language in [64] and the CPN/Beta simulator in [110]. Colour set restrictions have a number of similarities with the projections described in [40] and the colour simplifications in [21].

Exercises

Exercise 6.1.
Consider the resource allocation system from Sect. 1.2.

(a) Make a simulation of the timed CP-net in Fig. 6.13. This makes sense only if you have access to a tool which supports timed CP-nets.

(b) Implement the bar chart in Fig. 6.15. This makes sense only if you have access to a tool which supports code segments and reporting facilities.

Exercise 6.2.
Consider the ring protocol from Sect. 3.1.

(a) Construct a timed CP-net for the ring protocol (using an appropriate set of time delays).

(b) Make a simulation of the timed CP-net constructed in (a). This makes sense only if you have access to a tool which supports timed CP-nets.

(c) Discuss how the different kinds of reporting facilities can be used to create a graphical representation of the system state (e.g., describing the progress of the various messages).

(d) Make a set of charts and a set of code segments which implement your proposals from (c). This makes sense only if you have access to a tool which supports code segments and reporting facilities.

Chapter 7

Industrial Applications of Coloured Petri Nets

This chapter describes four of the projects which have used hierarchical CP-nets for modelling and analysis. All the projects have worked with reasonably large models, typically 15–50 pages and 25–150 page instances. They have all used the CPN tools described in Chap. 6, and they have been carried out in industrial environments, where efficiency parameters such as turn-around time and the use of man-hours are key issues.

We only sketch the main ideas behind the projects and the most important conclusions. However, there exist much more detailed descriptions of the projects, the models and the conclusions. References to these can be found in the bibliographical remarks. The presentations in this chapter contain a number of rather complex CPN pages. They are included to give the reader an idea about the modelling style and the complexity of the models. However, it is not the intention that the user should be able to understand all the details of these pages.

The number of industrial applications of CP-nets has been rapidly growing over the last years, as the theory and the computer tools have been developed. It is important that the experiences from such practical applications are disseminated from group to group. Only in this way can we learn from each other and avoid repeating the same errors. Thus we encourage participants in CPN projects to report their experiences, even though this will cost some time.

Section 7.1 reports on a project where CP-nets were used to model one of the standard communication protocols for digital telephone networks. Section 7.2 deals with a project in which the design of a new VLSI chip was analysed, in order to verify the logical correctness of the chip before it was manufactured. Section 7.3 describes how CP-nets can be used in the area of command and control, to analyse a complex radar surveillance system. Section 7.4 deals with a project in which CP-nets and SADT diagrams were used to develop a new algorithm for the control of electronic transfers of money between banks. In this project CP-nets were used throughout the entire design, analysis and implementation of the algorithm. Finally, Sect. 7.5 summarizes some of the general observations which can be made from the four projects.

7.1 Communication Protocol

This project was carried out in cooperation with the engineers at a large telecommunications company, and it involved the modelling and simulation of selected parts of an existing protocol for digital telephone networks. The considered protocol was the **BRI protocol** (Basic Rate Interface) for ISDN telephone networks (Integrated Services Digital Network). The focus was on the network layer, i.e., the third level in the OSI model (Open Systems Interconnection).

The modelling started from an existing description in **SDL** (Specification and Description Language). SDL is one of the graphical specification languages used for international telecommunications standards. For more information about SDL and how it can be translated into high-level Petri nets, see the bibliographical remarks. The translation to a hierarchical CP-net and the debugging of this net by means of simulation took only 16 days – for the basic part of the network layer. The work was done by a single modeller, who had wide experience with the CPN tools, but no prior knowledge of communication protocols.

The produced model was presented to engineers at the participating telecommunications company. This was done by making a manual simulation of typical occurrence sequences, and by making a super-automatic simulation – with a graphical representation similar to Fig. 6.12. According to the engineers, who all have wide experience with modelling and analysis of telephone systems, the CPN model provided the most detailed executable model which they had seen for this kind of protocol.

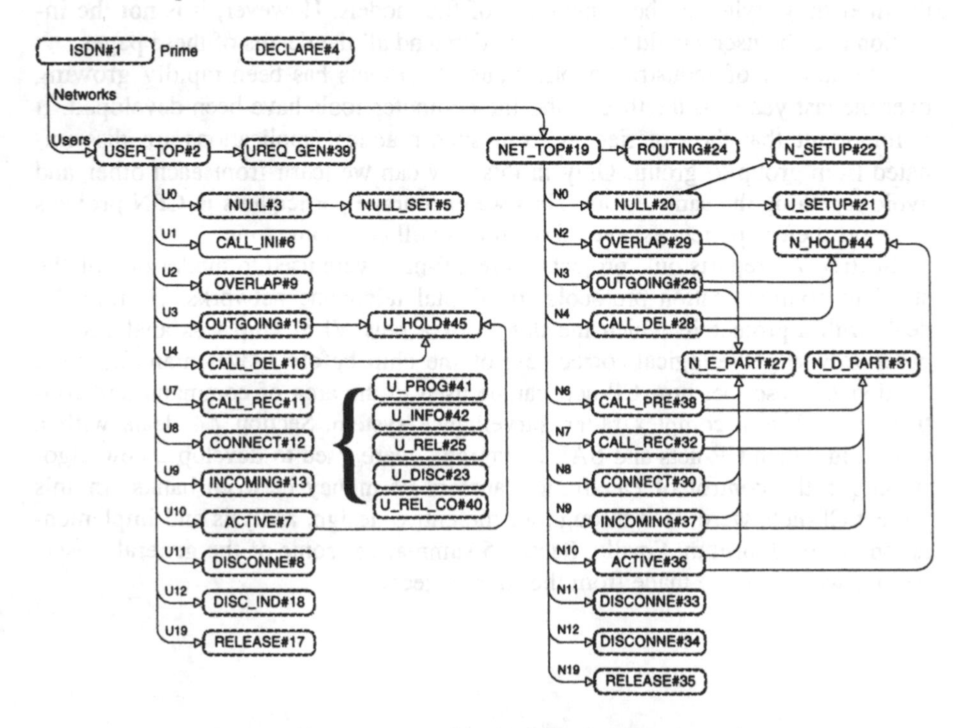

Fig. 7.1. Page hierarchy graph for the BRI protocol

Figure 7.1 shows the page hierarchy for the CPN diagram. All the page names have been truncated to at most eight characters. This convention keeps the page hierarchy graph readable (also in diagrams with very long page names). One of the next versions of the CPN tools will have a set of Name options allowing the modeller to specify, in more detail, how he wants names to be truncated.

The model has two parts. The 20 subpages of *USER_TOP#2* describe the actions of the user part, while the 18 subpages of *NET_TOP#19* describe the actions of the network part. Most of these 38 pages have a supernode which is called *Ui* or *Ni*, and this indicates that the page describes the actions which can happen when the user part is in state *Ui* or the network part in state *Ni*. The bracket in front of the five pages *U_PROG#41...U_REL_CO#40* indicates that they are subpages of all the pages in *NULL#3...RELEASE#17*. The five pages describe actions which are carried out in the same way in all user states, e.g., the *Disconnect* action. If the description of one of these actions has to be changed it is sufficient to modify one page of the CPN diagram, while in the SDL diagram it would be necessary to modify all those pages which use the action. *ROUTING#24* contains a representation of the services provided by the data link layer (i.e., the underlying protocol layer).

Later the modelling of a hold feature was included in the CPN diagram. This was done in a single day, and it turned out that it could be done by adding two extra pages, *U_HOLD#45* and *N_HOLD#44*, and by making a simple modification of the existing pages. A product colour set was given an additional component representing the hold status, and this meant that all arc expressions of the corresponding type had to be changed from a triple to a quadruple. This was

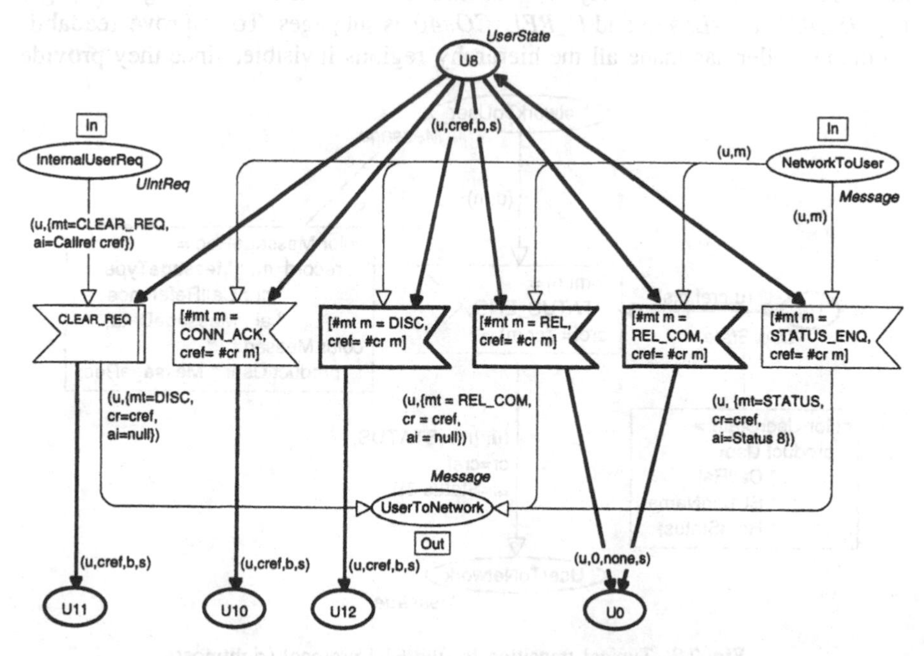

Fig. 7.2. Typical page for the BRI protocol *(CONNECT#12)*

trivial, however, since the added component, in the existing part of the diagram, was always passed from input tokens to output tokens without any modification. The simplicity by which the hold feature could be added to the CPN diagram was in sharp contrast to the original SDL model, where the inclusion made it necessary to duplicate the entire model, i.e., include many new pages. The same is true for three other features, which were not modelled in the CPN project, but could have been handled in a similar way. Obviously, it is much easier to maintain the CPN diagram, because it is sufficient to make modifications to one page instead of five. Moreover, it is easier to understand the individual features, because the description of each of them is confined to a few additional pages, instead of being spread out, as modifications of all the existing pages.

A typical representative of the *Ui* and *Ni* pages is shown in Fig. 7.2. The shapes of the transitions are carried over from the SDL diagram, where they have a formal meaning. In the CPN diagram the shapes have no formal meaning, but they make the diagram more readable for engineers who have experience with SDL. To improve readability the modeller has made some of the arc expressions invisible (when these are identical for neighbouring arcs).

From the page in Fig. 7.2 we see that the user part in state *U8* allows six possible actions to take place. When there is an *InternalUserRequest* the leftmost transition can occur. It creates a *UserToNetwork* message and the new user state becomes *U11*. When there is a *NetworkToUser* message one of the five remaining transitions can occur – the guards determine which one. Two of these transitions create a *NetworkToUser* message, and the new user state becomes either *U10, U12, U0* or remains *U8*. Three of the transitions are drawn with thick borders. This indicates that they are substitution transitions – having the pages *U_DISC#23, U_REL#25* and *U_REL_CO#40* as subpages. To improve readability the modeller has made all the hierarchy regions invisible, since they provide

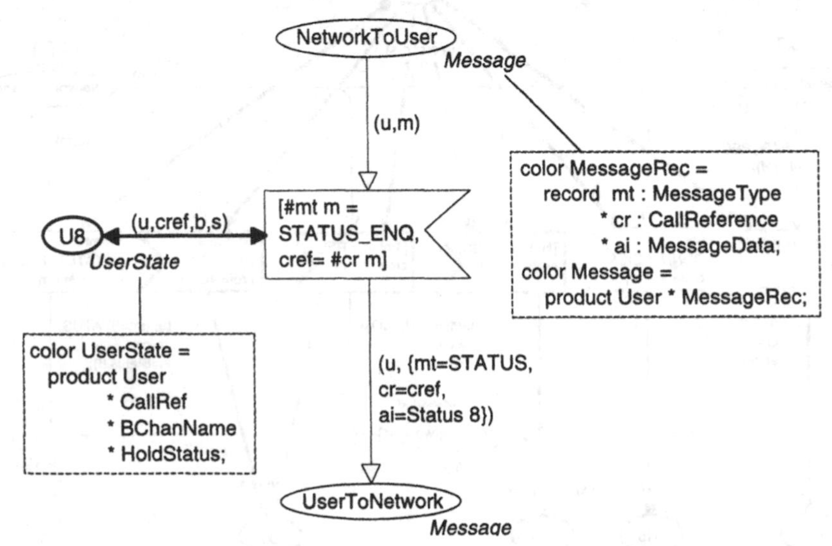

Fig. 7.3. Typical transition for the BRI protocol (rightmost transition of *CONNECT#12*)

no information which cannot be seen from the other net inscriptions surrounding the substitution transitions. It should also be noted that a global fusion set is used to fuse all the *U0* places together, and analogously there is a global fusion set for each of the other 23 kinds of *Ui* and *Ni* places. To improve readability the modeller has made all the fusion regions invisible.

A typical representative of a transition is shown in Fig. 7.3, where we take a closer look at the rightmost transition in Fig. 7.2. The dashed boxes show the declarations of the colour sets. These declarations are made in the global declaration node – together with all the other declarations of the CP-net. The guard specifies that the *MessageType* of the incoming message must be *STATUS_ENQ*, while the *CallReference* must match that of the user part. When the transition occurs, the user part remains in state *U8* and a *UserToNetwork* message is created. The new message has the same *User* and the same *CallReference* as the received message. It has *STATUS* as *MessageType* and *Status 8* as *MessageData*.

7.2 Hardware Chip

This project was carried out in cooperation with a large company which is a manufacturer of supercomputers. The purpose of the project was to investigate whether the use of CP-nets is able to speed up the design and validation of new VLSI chips, at the register transfer level.

Let us first describe how the design and validation of VLSI chips are carried out without the use of CP-nets. The chip designers specify a new chip by drawing a set of **block diagrams**, which each contains a set of nodes called blocks and the connections between them. Each block represents a functional unit with a specified input/output behaviour. A complex block may be described by means of a separate block diagram, which is related to the block in a way which is analogous to the relation between a substitution subpage and a supernode (in a CPN model). When the designers have finished a new chip, the block diagrams are translated, by a manual process, into a simulation program written in a special-purpose dialect of C. The simulation program is then executed on a large number of test data, typically 10 –20,000, and the output is analysed to detect any malfunctions. The design and validation strategy described above has a number of deficiencies – we return to these later.

Now let us describe the alternative design and validation strategy, involving CP-nets. The basic idea is to replace the manual translation, from the block diagrams into the C program, with an automatic translation into a CPN model. It is important to understand that it is *not* the intention to stop using block diagrams. The designers will still specify the VLSI chip by means of block diagrams, and they will follow the simulation of the CPN model by watching the block diagrams. To support the new strategy three things are needed:

- The existing editor tool for the block diagrams must be modified, so that it gets a fixed syntax with a well-defined semantics.
- It must be possible to translate a set of block diagrams into a hierarchical CP-net.

- The CPN simulator must be powerful enough to handle the rather complex VLSI designs, and efficient enough to make it possible to check the large number of test data.

The project only dealt with the last two issues, which were considered to be the most difficult. It was shown that the block diagrams could be translated into hierarchical CP-nets. This was done manually, but the translation process is rather straightforward, and we see no problems in implementing an automatic translation. The obtained CP-net only contains 15 pages, but during a simulation there are almost 150 page instances. This is because many subpages are used several times. This is the case, e.g., for a subpage which represents a 16-bit adder. The CP-net was simulated using the CPN simulator. When maximal graphical feedback was used, the simulation was slow – due to the many graphical objects which had to be updated in each step. However, when a more selective feedback was used, the speed became more reasonable.

Figure 7.4 shows a subpage from the CPN diagram. From this page it can be seen that the VLSI chip has a pipe-lined design with six different stages, represented by the substitution transitions *Stage1...Stage6*. Many of the places represent a 16-bit integer, being transferred from one stage to another. These places have a colour set which is of type *I16*. Other places represent clock signals sent from one stage to the previous stage, to indicate that data can be transferred. These places have surrounding arcs with arc expression c. Their colour set only contains one value, and the colour set region is invisible.

Each stage is modelled on a separate subpage and two of the more complex pages are shown in Fig. 7.5. The eight transitions in the middle of *Stage1* are all substitution transitions, and they have the same subpage. This represents the fact that the same functional unit is used eight times on the chip. In *Stage2* the four transitions *SUM1L, SUM1R, SUM2L* and *SUM2R* represent registers. These registers establish the border between *Stage2* and *Stage3*, and it can be seen that these transitions can only occur when they receive a clock signal from *Stage3*, via the two c-transitions in the rightmost part of *Stage2*. All the remaining transitions in *Stage2* are substitution transitions. *OR3* and *OR4* denote or-gates with three and four input signals, while + denotes 16-bit adders.

Now let us compare the new design and validation strategy with the old. First of all, it is easier to translate the block diagrams into a CP-net than it is to translate them into a C program. According to the participating company, the latter often takes several man-months, while the construction of the CP-net only took a few man-weeks. The translation is also more transparent, in the sense that it is much easier to recognize those parts of the CP-net which model a given block than it is to find the corresponding parts in the C program. This is due to the fact that each page of the CP-net has almost the same graphical layout as the corresponding block diagram. It means that it is relatively easy to change the CP-net to reflect any changes in the chip design. According to the chip manufacturer, it is a major problem to maintain the C program. Moreover, as stated above, we believe that it will be easy to automate the translation from block diagrams to CP-nets. This means that the modified CP-net can be obtained without any manual work at all.

Secondly, when the new strategy is fully implemented, the designers will be able to make simulations during the design process. This means that the knowledge and understanding which is acquired during the simulation can be used to improve the design itself, in a much more direct way than in the old strategy – where the entire validation is performed after the design has been finished.

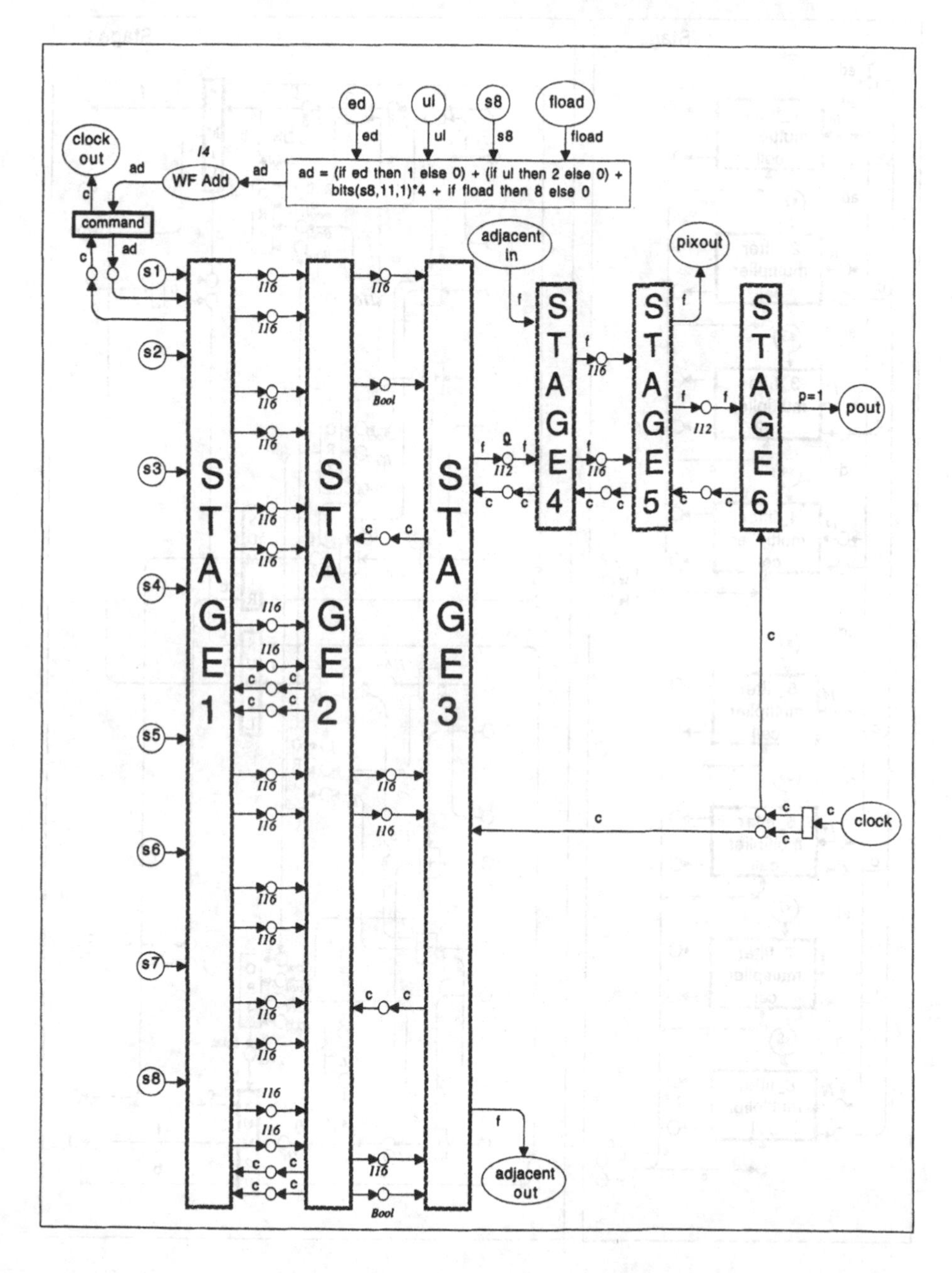

Fig. 7.4. Page from the VLSI chip

Thirdly, the validation techniques of the old strategy concentrate on logic correctness, i.e., the functionality of the VLSI chip. Very little concern (and no tests) seems to be given to those design decisions which deal with timing issues, e.g., the division into stages and the determination of an adequate clock rate. This is surprising, because the timing issues are crucial for the correct behaviour

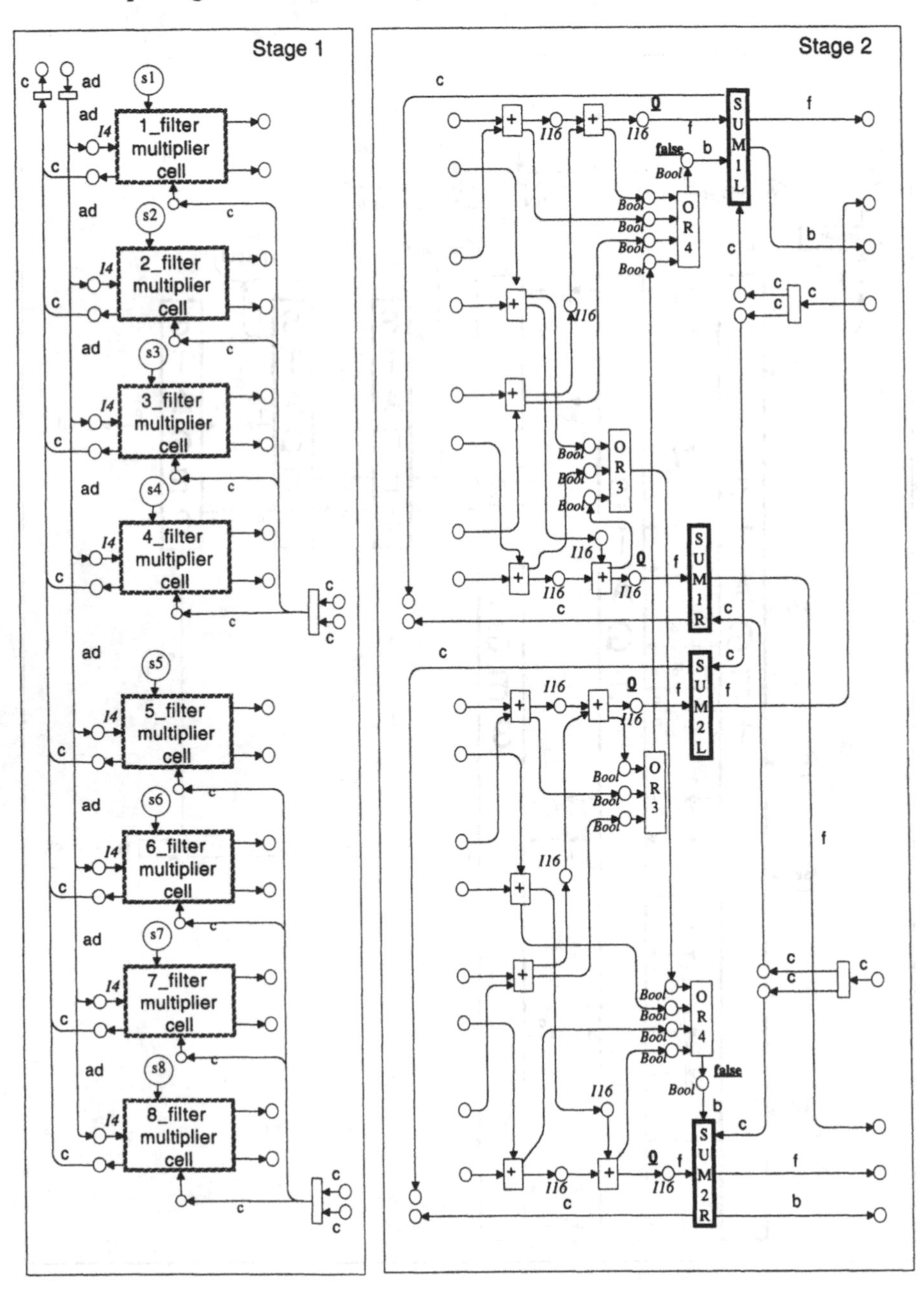

Fig. 7.5. Two subpages of the page in Fig. 7.4 (*Stage1* and *Stage2*)

and the effectiveness of the chip. Too-fast clocking implies malfunctioning while too-slow clocking implies unnecessary loss of speed. By means of timed CP-nets, it is rather straightforward to validate both the logic correctness and the timing issues by means of a single CPN model. However, the project was carried out before the time extensions were developed and implemented, and thus the project did not involve a validation of the time issues.

The only real drawback of the CPN approach was the speed of computation. It turned out that the execution of the C program was much faster than the execution of the CPN model. The CPN simulator at that time was simply unable to make the usual amount of test runs. However, it should be noted that the project was carried out immediately after the first version of the CPN simulator had been released. Based on the experience with this and other large models we have now improved the speed of the CPN simulator by more than a factor of 10. In addition to this we have got more powerful machines and super-automatic mode has been provided. Together this means that we are now in a situation where it makes sense also to deal with large sets of test data.

7.3 Radar Surveillance

This project was carried out in cooperation with AAMRL (Armstrong Aerospace Medical Research Laboratory) and it involved the modelling of a command post in the NORAD system (North American Radar Defense). The responsibility of the command post is to recommend different military actions, based upon an assessment of the rapidly changing status of the surveillance networks, defensive weapons and air traffic information. To do this the individual crew member communicates with many different types of equipment, other control posts and other members of the crew. There is a complex set of detailed rules saying what he must do in the different types of situations. The entire system can be compared to a very complex communication protocol, although a large part of the communication is between human beings and not between computers. The proper design of command posts, including procedures, equipment and staffing, is an on-going process, which is typical of the Command and Control area.

The purpose of the project was to use CP-nets to get an executable model of the command post, and use this model to get a better understanding of the command post, in order to improve its effectiveness and robustness. It was never the intention to use the CPN tools directly in the surveillance operations.

A team of modellers working at AAMRL created a description of the command post, by means of a graphical specification language called **SADT** (Structured Analysis and Design Technique). This language is in widespread use in some European countries and in the United States (where it is known as **IDEF**). The constructed SADT description was then augmented with more precise behavioural information, and the augmented model was automatically translated into a CP-net and simulated by means of the CPN simulator. For more details, see below. According to the people at AAMRL, the simulation gave an improved understanding of the command post, and they are now continuing the project modelling other parts of the NORAD system. It is the plan to model a number of command posts, and run the submodel for each of these on a separate machine, with its own copy of the CPN simulator. The submodels will communicate by means of files which are manipulated by code segments. This will be

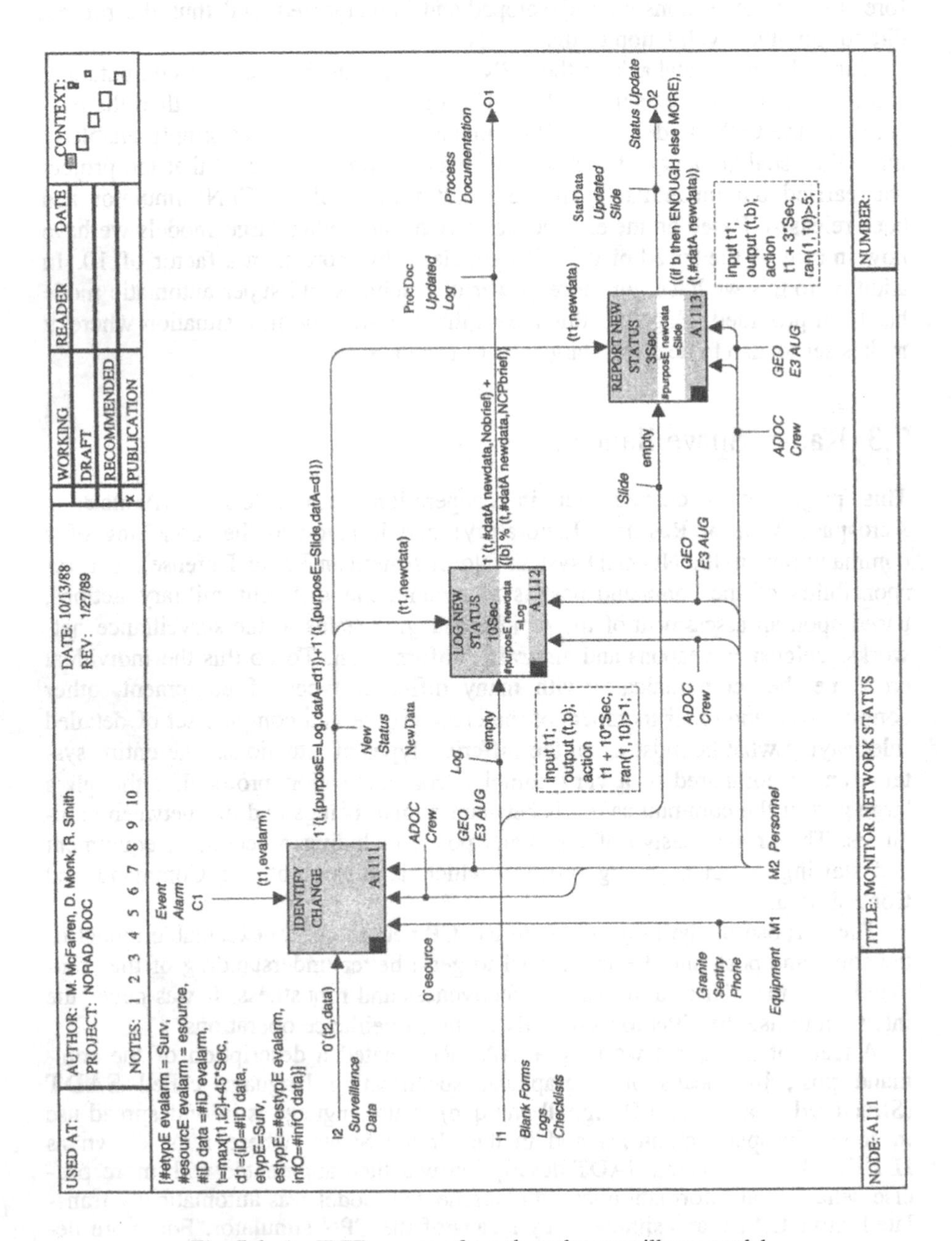

Fig. 7.6. An $\mathrm{IDEF_{CPN}}$ page from the radar surveillance model

similar to the way in which the real command posts communicate with each other via electronic networks.

SADT diagrams are in many respects similar to CP-nets, and this means that they consist of a set of pages. In the SADT terminology each page is called a dia-

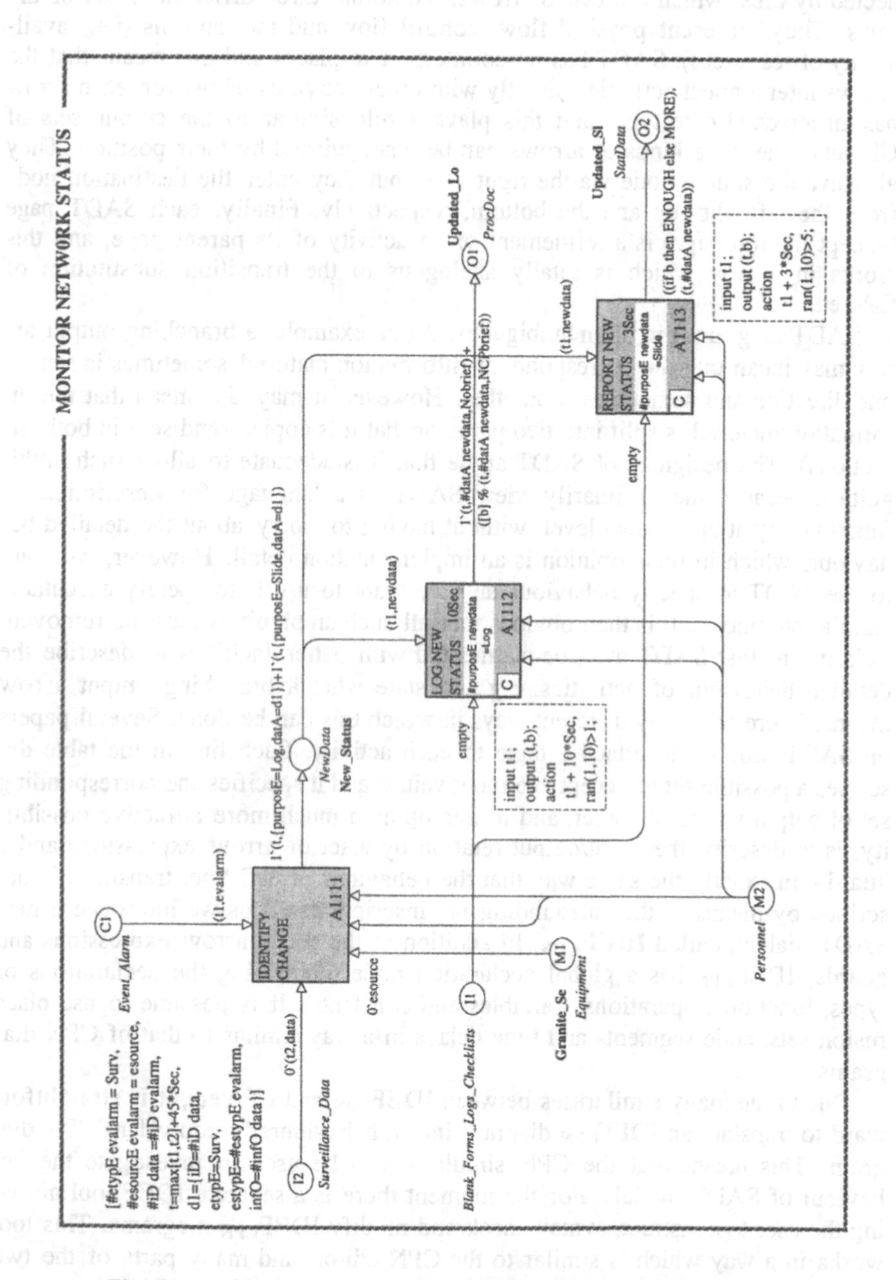

Fig. 7.7. The CPN page obtained from the IDEF$_{CPN}$ page in Fig. 7.6

gram. However, here we shall stick to the CPN convention and use the term diagram for a *set of pages* which constitutes a model. Each SADT page contains a number of rectangular boxes. They are called activities and they model actions in a way which is similar to the transitions of a CP-net. The activities are interconnected by arcs, which are called arrows. There are three different kinds of arrows. They represent physical flow, control flow and mechanisms (i.e., availability of resources). SADT has no counterpart to places and this means that the arrows interconnect activities directly with other activities. However, each arrow has an attached date type, and this plays a role similar to the colour sets of CP-nets. The three kinds of arrows can be distinguished by their position. They all leave the source node via the right side, but they enter the destination node from the left, the top and the bottom, respectively. Finally, each SADT page (except the top page) is a refinement of an activity of its parent page, and this works in a way which is totally analogous to the transition substitution of CP-nets.

SADT diagrams are often ambiguous. As an example, a branching output arrow may mean that the corresponding information/material sometimes is sent in one direction and sometimes in another. However, it may also mean that the information/material is split into two parts, or that it is copied (and sent in both directions). The designers of SADT argue that it is adequate to allow such ambiguities, because they primarily view SADT as a language for description of functionality at an abstract level, without having to worry about the detailed behaviour, which in their opinion is an implementation detail. However, we want to use SADT to specify behaviour and we want to use it to specify executable simulation models. It is then obvious that all such ambiguities must be removed. This means that SADT must be augmented with better facilities to describe the detailed behaviour of activities, e.g., to state what a branching output arrow means. There are many different ways in which this can be done. Several papers on SADT propose to attach a table to each activity. Each line in the table describes a possible set of acceptable input values and it specifies the corresponding set of output values. Another, and in our opinion much more attractive possibility, is to describe the input/output relation by a set of arrow expressions and a guard – in exactly the same way that the behaviour of a CP-net transition is described by means of the surrounding net inscriptions. Thus we introduce a new SADT dialect, called **IDEF$_{CPN}$**. In addition to the added arrow expressions and guards, IDEF$_{CPN}$ has a global declaration node (containing the declarations of types, functions, operations, variables and constants). It is possible to use place fusion sets, code segments and time delays in a way similar to that of CPN diagrams.

Due to the many similarities between IDEF$_{CPN}$ and CP-nets, it is straightforward to translate an IDEF$_{CPN}$ diagram into a behaviourally equivalent CPN diagram. This means that the CPN simulator can be used to investigate the behaviour of SADT models. For the moment there is a separate IDEF tool allowing the user to construct, syntax check and modify IDEF$_{CPN}$ diagrams. This tool works in a way which is similar to the CPN editor, and many parts of the two user interfaces are identical. The IDEF tool can create a file containing a textual representation of the IDEF$_{CPN}$ diagram, and this file can be read into the CPN simulator, where it is interpreted as a CPN diagram. The translation from $IDEF_{CPN}$ to CPN diagrams is thus totally automatic. Later, it is the plan to integrate a copy of the CPN simulator into the IDEF tool itself. This will mean that the turn-around time will be faster, because it is then possible to edit and simulate in the same tool. Such an integration will also mean that the user will see the simulation results directly on the $IDEF_{CPN}$ diagram. For the moment he views the results on the corresponding CPN diagram – but this is not a big problem because the two diagrams look almost identical (except that the former does not have places). Figure 7.6 shows an $IDEF_{CPN}$ page from the radar surveillance system, and Fig. 7.7 shows the corresponding CPN page, as it is obtained by the automatic translation.

7.4 Electronic Funds Transfer

This project was carried out in cooperation with two banks, Société Générale and Marine Midland Bank of New York. The project involved the design and implementation of software to *control* the electronic transfer of money between banks. The speed of modern bank operations means that banks often make commitments which are based on money which they do not have, but expect to receive within the next few minutes. What happens if this money is delayed or never arrives?

Two managers at the involved banks had an idea for a new control strategy, which would allow the responsible staff to use computer support to control the funds transfer. Until now all the decisions about acceptance or rejection of the individual transactions have been made manually – although the transactions themselves are performed via special computer networks. The two managers concretized their idea in terms of a relatively small **SADT diagram** which was created by means of the **IDEF tool** and contained a rather informal description of the proposed algorithm. For an introduction to SADT diagrams and the IDEF tool, see Sect. 7.3. The constructed SADT diagram was translated to a CP-net and more accurate behavioural information was added, by an experienced CPN modeller. This was done in close cooperation with the two bank managers, who also participated in the debugging, during which the original algorithm was tested and slightly improved. The additional behavioural information could just as well have been added before the translation, i.e., by means of the IDEF tool instead of the CPN editor.

During the project there were several different versions of the CPN model. The first of these was obtained more or less directly from the SADT diagram, and it was rather crude – with simple arc expressions and very simple colour sets. This model was primarily used to describe the flow of data, while the actual manipulations of the data were ignored. Later, the arc expressions were made more precise, a large number of complex data types were declared (and used as colour sets), complex CPN ML functions were declared (e.g., to search, sort and merge files), and finally most of the behavioural information was moved to code segments. In the final CPN model most transitions have arc expressions which consist of a single variable, and complex code segments determining the values of the output variables from the values of the input variables. It took 5 man-weeks to create the SADT diagram, only 1 man-week to get the first CPN diagram, and 16 man-weeks to develop this into the final CPN model.

In the first part of the project the graphical interface of the tools was of very high importance, and it was the graphical aspects of SADT and CP-nets that made it possible for the bank managers to concretize their ideas. However, later it turned out that the graphical interface became less important, while the output files produced by the simulations became more important, and hence the project started to use super-automatic simulations. A simulation works with a number of input files, describing transfers which have already been made that day and transfers which are registered but not yet executed. From these input files, which typically contain 25–50,000 rather complex records, a number of output files are produced. This takes 5–10 minutes, and from these output files the bank staff determines the control strategy to be used for the next 15–20 minutes, at which time a new set of simulation results is ready.

In this project the CPN tools were used as a **case tool**. The new strategy was specified by means of the IDEF tool and the CPN editor. Then it was validated by means of the CPN simulator. Finally, the SML code (produced by the CPN simulator) was used as a stand-alone program.

The new control strategy, proposed by the two bank managers, seems to be working as expected. The strategy is being tested on historical bank data, using the SML code produced by the CPN simulator. When these tests are finished, it will be determined whether the project will continue. If it is continued, the CPN model (and the IDEF_{CPN} model) will be extended to reflect additional aspects of funds transfer. Moreover, a graphical user interface will be added, allowing the bank staff to interact with the model in a more natural and straightforward way – without knowing anything about CP-nets. This user interface will be created by letting the code segments use the graphical routines of the CPN tools. These routines are also available in the stand-alone SML environment, and thus it will still be possible to obtain the final SML code automatically from the CPN simulator.

7.5 General Observations About CPN Applications

From the applications reported in this chapter and some of the applications mentioned in the bibliographical remarks two important observations can be made.

First of all, it is often adequate to use CPN models in connection with different front-end languages, e.g., SDL, SADT and block diagrams. The reason may be that there already exist descriptions in these languages, or that the projects involve people who are familiar with some of the languages and thus prefer to use them instead of learning a totally new formalism. It will also sometimes be sensible to make a tailored language, with a semantics based on CP-nets, but a syntax adapted to the problem area. This is for instance done by the designers of the Vista language, who have defined the semantics of their graphical specification language for distributed software by specifying a translation to CP-nets. For more information see the bibliographical remarks.

Secondly, it is often the case that the graphical representation, which is very important in the early design and validation phases, later becomes less interesting. In this case the modellers may turn to super-automatic simulation. The results obtained from this kind of simulation are very similar to those which may be obtained from building a prototype. The CPN tools can hence be used as a case tool. In certain application areas, it is even possible to use the resulting SML code as the final implementation. This will be even more attractive when it becomes possible to write code segments in different languages, e.g., C++, Pascal and Prolog (as mentioned in Sect. 6.2).

Bibliographical Remarks

Detailed presentations of the CPN projects presented in this chapter can be found in [50] (BRI protocol), [100] (VLSI chip), [101] (radar surveillance) and [87] (electronic funds transfer). A description of SDL and its translation to high-level nets can be found in [27], [68] and [99], while a description of SADT can be found in [70] and a description of the Vista language in [60].

CP-nets and other kinds of high-level Petri nets are in use within a large variety of different areas. For more information about some of these application areas, see [10], [22], [25], [29], [32], [37], [43] (communication protocols), [2] (computer architecture), [112] (computer organization), [117] (data bases), [3], [46], [98] (distributed algorithms), [71], [104] (flexible manufacturing systems), [81] (human-machine interaction), [78] (inconsistency tolerance), [75], [83] (logic programs), [118] (office automation), [77] (robotics), [53] (semantics of programming languages) and [4], [28], [47] (software development methods).

References

1. K. Albert, K. Jensen, R.M. Shapiro: *Design/CPN. A Tool Package Supporting the Use of Coloured Petri Nets.* Petri Net Newsletter no. 32 (April 1989), Special Interest Group on Petri Nets and Related System Models, Gesellschaft für Informatik (GI), Germany, 1989, 22–36.

2. J.L. Baer: *Modelling Architectural Features with Petri Nets.* In: W. Brauer, W. Reisig and G. Rozenberg (eds.): Petri Nets: Application and Relationships to Other Models of Concurrency, Advances in Petri Nets 1986 Part II, Lecture Notes in Computer Science Vol. 255, Springer-Verlag 1987, 258–277.

3. G. Balbo, S.C. Bruell, P. Chen, G. Chiola: *An Example of Modelling and Evaluation of a Concurrent Program Using Colored Stochastic Petri Nets: Lamport's Fast Mutual Exclusion Algorithm.* IEEE Transactions on Parallel and Distributed Systems, 3 (1992). Also in [59], 533–559.

4. M. Baldassari, G. Bruno: *PROTOB: An Object Oriented Methodology for Developing Discrete Event Dynamic Systems.* Computer Languages 16 (1991), Pergamon Press, 39–63. Also in [59], 624–648.

5. E. Battiston, F. De Cindio, G. Mauri: *OBJSA Nets: A Class of High-level Nets Having Objects as Domains.* In: G. Rozenberg (ed.): Advances in Petri Nets 1988, Lecture Notes in Computer Science Vol. 340, Springer-Verlag 1988, 20–43. Also in [59], 189–212.

6. B. Baumgarten: *Petri-Netze. Grundlagen und Anwendungen.* Wissenschaftsverlag, Mannheim, 1990.

7. E. Best, C. Fernandez: *Notations and Terminology of Petri Net Theory.* Petri Net Newsletter no. 23 (April 1986), Special Interest Group on Petri Nets and Related System Models, Gesellschaft für Informatik (GI), Germany, 1986, 21–46.

8. E. Best, C. Fernandez: *Nonsequential Processes.* EATCS Monographs on Theoretical Computer Science, Vol. 13, Springer-Verlag, 1988.

9. J. Billington: *Many-sorted High-level Nets.* Proceedings of the Third International Workshop on Petri Nets and Performance Models, Kyoto 1989, 166–179. Also in [59], 123–136.

10. J. Billington, G. Wheeler, M. Wilbur-Ham: *Protean: A High-level Petri Net Tool for the Specification and Verification of Communication Protocols.* IEEE Transactions on Software Engineering, Special Issue on Tools for Computer Communication Systems 14 (1988), 301–316. Also in [59], 560–575.

11. G.W. Brams: *Réseaux de Petri: Théorie et Pratique.* Tome 1: Théorie et Analyse, Tome 2: Modélisation et Applications, Edition Masson, 1983.

12. G.W. Brams: *Le Reti di Petri: Teoria e Pratica.* Tomo 1: Teoria e Analisi, Tomo 2: Modellazione e Applicazioni, Italian translation of [11], Edition Masson, 1985.

13. G.W. Brams: *Las Redes de Petri: Teoria y Practica.* Tomo 1: Teoria y Analisis, Tomo 2: Modelizacion y Applicaciones, Spanish translation of [11], Edition Masson, 1986.

14. W. Brauer (ed.): *Net Theory and Applications.* Proceedings of the Advanced Course on General Net Theory of Processes and Systems, Hamburg 1979, Lecture Notes in Computer Science Vol. 84, Springer-Verlag 1980.

15. W. Brauer, W. Reisig, G. Rozenberg (eds.): *Petri Nets: Central Models and Their Properties.* Advances in Petri Nets 1986 Part I, Lecture Notes in Computer Science Vol. 254, Springer-Verlag 1987.

16. W. Brauer, W. Reisig, G. Rozenberg (eds.): *Petri Nets: Applications and Relationships to Other Models of Concurrency.* Advances in Petri Nets 1986 Part II, Lecture Notes in Computer Science Vol. 255, Springer-Verlag 1987.

17. J.A. Carrasco: *Automated Construction of Compound Markov Chains from Generalized Stochastic High-level Petri Nets.* Proceedings of the Third International Workshop on Petri Nets and Performance Models, Kyoto 1989, 93–102. Also in [59], 494 –503.

18. G. Chehaibar: *Use of Reentrant Nets in Modular Analysis of Coloured Petri Nets.* In: G. Rozenberg (ed.): Advances in Petri Nets 1991, Lecture Notes in Computer Science Vol. 524, Springer-Verlag 1991, 58–77. Also in [59], 596 – 617.

19. G. Chiola, C. Dutheillet, G. Franceschinis, S. Haddad: *On Well-Formed Coloured Nets and Their Symbolic Reachability Graph.* In [59], 373–396.

20. G. Chiola, C. Dutheillet, G. Franceschinis, S. Haddad: *Stochastic Well-Formed Coloured Nets and Multiprocessor Modelling Applications.* In [59], 504 –530.

21. G. Chiola, G. Franceschinis: *A Structural Colour Simplification in Well-Formed Coloured Nets.* In PNPM91: Petri Nets and Performance Models. Proceedings of the 4th International Workshop, Melbourne, Australia 1991, IEEE Computer Society Press, 144 –153.

22. S. Christensen, L.O. Jepsen: *Modelling and Simulation of a Network Management System Using Hierarchical Coloured Petri Nets.* In: E. Mosekilde (ed.): Modelling and Simulation 1991. Proceedings of the 1991 European Simulation Multiconference, Copenhagen, 1991, Society for Computer Simulation 1991, 47–52.

23. S. Christensen, L. Petrucci: *Towards a Modular Analysis of Coloured Petri Nets.* In: K. Jensen (ed.): Application and Theory of Petri Nets 1992. Proceedings of the 13th International Petri Net Conference, Sheffield 1992, Lecture Notes in Computer Science Vol. 616, Springer-Verlag 1992,113–133.

24. P. Chrzastowski-Wachtel: *A Proposition for Generalization of Liveness and Fairness Properties.* Proceedings of the 8th European Workshop on Application and Theory of Petri Nets, Zaragoza 1987, 215–236.

25. B. Cousin, J.M. Couvreur, C Dutheillet, P. Estraillier: *Validation of a Protocol Managing a Multi-token Ring Architecture.* Proceedings of the 9th European Workshop on Application and Theory of Petri Nets, Vol. II, Venice 1988.

26. J.M. Couvreur, J. Martínez: *Linear Invariants in Commutative High Level Nets.* In: G. Rozenberg (ed.): Advances in Petri Nets 1990, Lecture Notes in Computer Science Vol. 483, Springer-Verlag 1991, 146–165. Also in [59], 284 –302.

27. F. De Cindio, G. Lanzarone, A. Torgano: *A Petri Net Model of SDL.* Proceedings of the 5th European Workshop on Application and Theory of Petri Nets, Aarhus 1984, 272–289.

28. R. Di Giovanni: *Hood Nets.* In: G. Rozenberg (ed.): Advances in Petri Nets 1991, Lecture Notes in Computer Science Vol. 524, Springer-Verlag 1991, 140 –160.

29. M. Diaz: *Petri Net Based Models in the Specification and Verification of Protocols.* In: W. Brauer, W. Reisig and G. Rozenberg (eds.): Petri Nets: Applications and Relationships to Other Models of Concurrency, Advances in Petri Nets 1986 Part II, Lecture Notes in Computer Science Vol. 255, Springer-Verlag 1987, 135–170.

30. C. Dimitrovici, U. Hummert, L. Petrucci: *Semantics, Composition and Net Properties of Algebraic High-level Nets.* In: G. Rozenberg (ed.): Advances in Petri Nets 1991, Lecture Notes in Computer Science Vol. 524, Springer-Verlag 1991, 93–117.

31. C. Dutheillet, S. Haddad: *Regular Stochastic Petri Nets.* In: G. Rozenberg (ed.): Advances in Petri Nets 1990, Lecture Notes in Computer Science Vol. 483, Springer-Verlag 1991, 186–210. Also in [59], 470 – 493.

32. P. Estraillier, C. Girault: *Petri Nets Specification of Virtual Ring Protocols.* In: A. Pagnoni, G. Rozenberg (eds.): Applications and Theory of Petri Nets, Informatik-Fachberichte Vol. 66, Springer-Verlag 1983, 74 –85.

33. R. Fehling: *A Concept for Hierarchical Petri Nets with Building Blocks.* Proceedings of the 12th International Conference on Application and Theory of Petri Nets, Aarhus 1991, 370 –389.

34. F. Feldbrugge: *Petri Net Tool Overview 1989.* In: G. Rozenberg (ed.): Advances in Petri Nets 1989, Lecture Notes in Computer Science Vol. 424, Springer-Verlag 1990, 151–178.

35. F. Feldbrugge, K. Jensen: *Computer Tools for High-level Petri Nets.* In [59], 691–717.

36. A. Finkel: *The Minimal Coverability Graph for Petri Nets.* In: G. Rozenberg (ed.): Advances in Petri Nets 1993, Lecture Notes in Computer Science Vol. 674, Springer-Verlag 1993, 210 –234.

37. G. Florin, C. Kaiser, S. Natkin: *Petri Net Models of a Distributed Election Protocol on Undirectional Ring.* Proceedings of the 10th International Conference on Application and Theory of Petri Nets, Bonn 1989, 154–173.

38. H.J. Genrich, K. Lautenbach, P.S. Thiagarajan: *Elements of General Net Theory.* In: W. Brauer (ed.): Net theory and applications. Proceedings of the Advanced Course on General Net Theory of Processes and Systems, Hamburg 1979, Lecture Notes in Computer Science Vol. 84, Springer-Verlag 1980, 21–163.

39. H.J. Genrich, K. Lautenbach: *System Modelling with High-level Petri Nets.* Theoretical Computer Science 13 (1981), North-Holland, 109–136.

40. H.J. Genrich: *Projections of C/E-systems.* In: G. Rozenberg (ed.): Advances in Petri Nets 1985, Lecture Notes in Computer Science Vol. 222, Springer-Verlag 1986, 224–232.

41. H.J. Genrich: *Predicate/Transition Nets.* In: W. Brauer, W. Reisig, G. Rozenberg (eds.): Petri Nets: Central Models and Their Properties, Advances in Petri Nets 1986 Part I, Lecture Notes in Computer Science Vol. 254, Springer-Verlag 1987, 207–247. Also in [59], 3–43.

42. H.J. Genrich: *Equivalence Transformations of PrT-Nets.* In: G. Rozenberg (ed.): Advances in Petri Nets 1989, Lecture Notes in Computer Science, Vol. 424, Springer-Verlag 1990, 179–208. Also in [59], 426–455.

43. C. Girault, C. Chatelain, S. Haddad: *Specification and Properties of a Cache Coherence Protocol Model.* In: G. Rozenberg (ed.): Advances in Petri Nets 1987, Lecture Notes in Computer Science, Vol. 266, Springer-Verlag 1987, 1–20. Also in [59], 576–595.

44. S. Haddad: *A Reduction Theory for Coloured Nets.* In: G. Rozenberg (ed.): Advances in Petri Nets 1989, Lecture Notes in Computer Science, Vol. 424, Springer-Verlag 1990, 209–235. Also in [59], 399–425.

45. R. Harper: *Introduction to Standard ML.* Technical Report ECS-LFCS-86-14, University of Edinburgh, Department of Computer Science, 1986.

46. G. Hartung: *Programming a Closely Coupled Multiprocessor System with High Level Petri Nets.* In: G. Rozenberg (ed.): Advances in Petri Nets 1988, Lecture Notes in Computer Science Vol. 340, Springer-Verlag 1988, 154–174.

47. T. Hildebrand, H. Nieters, N Trèves: *The Suitability of Net-based Graspin Tools for Monetics Applications.* Proceedings of the 11th International Conference on Application and Theory of Petri Nets, Paris 1990, 139–160.

48. P. Huber, A.M. Jensen, L.O. Jepsen, K. Jensen: *Reachability Trees for High-level Petri Nets.* Theoretical Computer Science 45 (1986), North-Holland, 261–292. Also in [59], 319–350.

49. P. Huber, K. Jensen, R.M. Shapiro: *Hierarchies in Coloured Petri Nets.* In: G. Rozenberg (ed.): Advances in Petri Nets 1990, Lecture Notes in Computer Science Vol. 483, Springer-Verlag 1991, 313–341. Also in [59], 215–243.

50. P. Huber, V.O. Pinci: *A Formal Executable Specification of the ISDN Basic Rate Interface.* Proceedings of the 12th International Conference on Application and Theory of Petri Nets, Aarhus 1991, 1–21.

51. K. Jensen: *Coloured Petri Nets and the Invariant Method.* Theoretical Computer Science 14 (1981), North-Holland, 317–336.

52. K. Jensen: *High-level Petri Nets.* In: A. Pagnoni, G. Rozenberg (eds.): Applications and Theory of Petri Nets, Informatik-Fachberichte Vol. 66, Springer-Verlag 1983, 166–180.

53. K. Jensen, E.M. Schmidt: *Pascal Semantics by a Combination of Denotational Semantics and High-level Petri Nets.* In: G. Rozenberg (ed.): Advances in Petri Nets 1985, Lecture Notes in Computer Science Vol. 222, Springer-Verlag 1986, 297–329.

54. K. Jensen: *Coloured Petri Nets.* In: W. Brauer, W. Reisig, G. Rozenberg (eds.): Petri Nets: Central Models and Their Properties, Advances in Petri Nets 1986 Part I, Lecture Notes in Computer Science Vol. 254, Springer-Verlag 1987, 248–299.

55. K. Jensen, et. al: *Design/CPN. A Reference Manual.* Meta Software and Computer Science Department, University of Aarhus, Denmark. On-line version: http://www.daimi.aau.dk/designCPN/.

56. K. Jensen, et. al.: *Design/CPN Occurrence Graph Manual.* Meta Software and Computer Science Department, University of Aarhus, Denmark. On-line version: http://www.daimi.aau.dk/designCPN/.

57. K. Jensen, P. Huber: *Design/CPN Extensions.* Meta Software Corporation, 125 Cambridge Park Drive, Cambridge MA 02140, USA, 1990.

58. K. Jensen: *Coloured Petri Nets: A High-level Language for System Design and Analysis.* In: G. Rozenberg (ed.): Advances in Petri Nets 1990, Lecture Notes in Computer Science Vol. 483, Springer-Verlag 1991, 342–416. Also in [59], 44 –122.

59. K. Jensen, G. Rozenberg (eds.): *High-level Petri Nets. Theory and Application.* Springer-Verlag, 1991.

60. E. de Jong, M.R. van Steen: *Vista: A Specification Language for Parallel Software Design.* Proceedings of the 3rd International Workshop on Software Engineering and its Applications, Toulouse, 1990.

61. R.M. Karp, R.E. Miller: *Parallel Program Schemata.* Journal of Computer and System Sciences, Vol. 3, 1969, 147–195.

62. A. Karsenty: *Interactive Graphical Reporting Facilities for Design/CPN.* Master's Thesis, University of Paris Sud, Computer Science Department, 1990.

63. A. Kiehn: *Petri Net Systems and Their Closure Properties.* In: G. Rozenberg (ed.): Advances in Petri Nets 1989, Lecture Notes in Computer Science, Vol. 424, Springer-Verlag 1990, 306 –328.

64. B.B. Kristensen, O.L. Madsen, B. Møller-Pedersen, K. Nygaard: *The BETA Programming Language.* In: B.D. Shriver, P. Wegner (eds.): Research Directions in Object-Oriented Languages, MIT Press, 1987.

65. B. Krämer, H.W. Schmidt: *Types and Modules for Net Specifications.* In: K. Voss, H.J. Genrich, G. Rozenberg (eds.): Concurrency and Nets, Advances in Petri Nets, Springer-Verlag 1987, 269–286. Also in [59], 171–188.

66. D. Lehmann, A. Pnueli, J. Stavi: *Impartiality, Justice and Fairness: The Ethics of Concurrent Termination.* In: S. Even, O. Kariv (eds.): Automata, Languages and Programming, Proceedings of ICALP 1981, Lecture Notes in Computer Science Vol. 115, Springer-Verlag 1981, 264–277.

67. C. Lin, D.C. Marinescu: *Stochastic High-Level Petri Nets and Applications.* IEEE Transactions on Computers, 37 (1988), 815–825. Also in [59], 459–469.

68. M. Lindqvist: *Translation of the Specification Language SDL into Predicate/Transition Nets.* Licentiate's Thesis, Helsinki University of Technology, Digital Systems Laboratory, 1987.

69. M. Lindqvist: *Parameterized Reachability Trees for Predicate/Transition Nets.* In: G. Rozenberg (ed.): Advances in Petri Nets 1993, Lecture Notes in Computer Science Vol. 674, Springer-Verlag 1993, 301–324. Also in [59], 351–372.

70. D.A. Marca, C.L. McGowan: *SADT.* McGraw-Hill, New York, 1988.

71. J Martínez, P. Muro, M. Silva: *Modelling, Validation and Software Implementation of Production Systems using High Level Petri Nets.* Proceedings of the IEEE International Conference on Robotics and Automation, Raleigh (USA) 1987, 1180–1185. Also in [59], 618–623.

72. G. Memmi, J. Vautherin: *Analysing Nets by the Invariant Method.* In: W. Brauer, W. Reisig, G. Rozenberg (eds.): Petri Nets: Central Models and Their Properties, Advances in Petri Nets 1986 Part I, Lecture Notes in Computer Science Vol. 254, Springer-Verlag 1987, 300–336. Also in [59], 247–283.

73. R. Milner, R. Harper, M. Tofte: *The Definition of Standard ML.* MIT Press, 1990.

74. R. Milner, M. Tofte: *Commentary on Standard ML.* MIT Press, 1991.

75. T. Murata, D. Zhang: *A Predicate-Transition Net Model for Parallel Interpretation of Logic Programs.* IEEE Transactions on Software Engineering 14 (1988), 481–497.

76. T. Murata: *Petri Nets: Properties, Analysis and Applications.* Proceedings of the IEEE Vol. 77 no. 4 (April 1989), 541–580.

77. T. Murata, P.C. Nelson, J. Yim: *A Predicate-Transition Net Model for Multiple Agent Planning.* Information Sciences 57–58 (1991), 361–384.

78. T. Murata, V.S. Subrahmanian, T. Wakayama: *A Petri Net Model for Reasoning in the Presence of Inconsistency.* IEEE Transactions on Knowledge and Data Engineering 3 (1991), 281–292.

79. Y. Narahari: *On the Invariants of Coloured Petri Nets.* In: G. Rozenberg (ed.): Advances in Petri Nets 1985, Lecture Notes in Computer Science Vol. 222, Springer-Verlag 1986, 330–345.

80. H. Oberquelle: *Communication by Graphic Net Representations.* Bericht Nr. 75, Fachbereich Informatik, Universität Hamburg 1981.

81. H. Oberquelle: *Human-Machine Interaction and Role/Function/Action-Nets.* In: W. Brauer, W. Reisig, G. Rozenberg (eds.): Petri Nets: Applications and Relationships to Other Models of Concurrency, Advances in Petri Nets 1986 Part II, Lecture Notes in Computer Science Vol. 255, Springer-Verlag 1987, 171–190.

82. L. Paulson: *ML for the Working Programmer.* Cambridge University Press, 1991.

83. G. Peterka, T. Murata: *Proof Procedure and Answer Extraction in Petri Net Model of Logic Programs.* IEEE Transactions on Software Engineering 15 (1989), 209–217.

84. J.L. Peterson: *Petri Net Theory and the Modeling of Systems.* Prentice-Hall, 1981.

85. C.A. Petri: *Kommunikation mit Automaten.* Schriften des IIM Nr. 2, Institut für Instrumentelle Mathematik, Bonn, 1962. *English translation:* Technical Report RADC-TR-65-377, Griffiths Air Force Base, New York, Vol. 1, Suppl. 1, 1966.

86. L. Petrucci: *Combining Finkel's and Jensen's Reduction Techniques to Build Covering Trees for Coloured Nets.* Petri Net Newsletter no. 36 (August 1990), Special Interest Group on Petri Nets and Related System Models, Gesellschaft für Informatik (GI), Germany, 1990, 32–36.

87. V.O. Pinci, R.M. Shapiro: *An Integrated Software Development Methodology Based on Hierarchical Colored Petri Nets.* In: G. Rozenberg (ed.): Advances in Petri Nets 1991, Lecture Notes in Computer Science Vol. 524, Springer-Verlag 1991, 227–252. Also in [59], 649–667.

88. H. Plünnecke, W. Reisig: *Bibliography of Petri Nets.* In: G. Rozenberg (ed.): Advances in Petri Nets 1991, Lecture Notes in Computer Science Vol. 524, Springer-Verlag 1991, 317–572.

89. *PNPM89: Petri Nets and Performance Models.* Proceedings of the 3rd International Workshop, Kyoto Japan 1989, IEEE Computer Society Press.

90. *PNPM91: Petri Nets and Performance Models.* Proceedings of the 4th International Workshop, Melbourne, Australia 1991, IEEE Computer Society Press.

91. C. Reade: *Elements of Functional Programming.* International Computer Science Series, Addison-Wesley, 1989.

92. W. Reisig: *Petri Nets. An Introduction.* EATCS Monographs on Theoretical Computer Science, Vol. 4, Springer-Verlag, 1985.

93. W. Reisig: *Petrinetze. Eine Einführung.* Springer-Verlag, 1986.

94. W. Reisig: *Petri Nets with Individual Tokens.* Theoretical Computer Science 41 (1985), North-Holland, 185–213.

95. W. Reisig: *Petri Nets and Algebraic Specifications.* Theoretical Computer Science 80 (1991), North-Holland, 1–34. Also in [59], 137–170.

96. W. Reisig: *A Primer in Petri Net Design.* Springer-Verlag, 1992.

97. G. Rozenberg: *Behaviour of Elementary Net Systems.* In: W. Brauer, W. Reisig, G. Rozenberg (eds.): Petri Nets: Central Models and Their Properties, Advances in Petri Nets 1986 Part I, Lecture Notes in Computer Science Vol. 254, Springer-Verlag 1987, 60–94.

98. M. Rukoz, R. Sandoval: *Specification and Correctness of Distributed Algorithms by Coloured Petri Nets.* Proceedings of the 9th European Workshop on Application and Theory of Petri Nets, Vol. II, Venice 1988.

99. *Functional Specification and Description Language SDL.* In: CCITT Yellow Book, Vol. VI, recommendations Z.101–Z.104, CCITT, Geneva, 1981.

100. R.M. Shapiro: *Validation of a VLSI Chip Using Hierarchical Coloured Petri Nets.* Journal of Microelectronics and Reliability, Special Issue on Petri Nets, Pergamon Press, 1991. Also in [59], 667–687.

101. R.M. Shapiro, V.O. Pinci, R. Mameli: *Modelling a NORAD Command Post Using SADT and Coloured Petri Nets.* Proceedings of the IDEF Users Group, Washington DC, May 1990.

102. M. Silva, J. Martínez, P. Ladet, H. Alla: *Generalized Inverses and the Calculation of Symbolic Invariants for Coloured Petri Nets.* Technique et Science Informatiques 4 (1985), 113–126. Also in [59], 303–315.

103. M. Silva: *Las Redes de Petri: En la Automática y la Informática.* Editorial AC, Madrid, 1985.

104. M. Silva, R. Valette: *Petri Nets and Flexible Manufacturing.* In: G. Rozenberg (ed.): Advances in Petri Nets 1989, Lecture Notes in Computer Science, Vol. 424, Springer-Verlag 1990, 374–417.

105. S. Sokolowski: *Applicative High Order Programming; the Standard ML Perspective.* Chapman and Hall, 1991.

106. P. Starke: *Analyse von Petri Netz Modellen.* Teubner, Stuttgart, 1990.

107. F.J.W. Symons: *Modelling and Analysis of Communication Protocols Using Numerical Petri Nets.* Ph.D. dissertation, Report 152, Department of Electrical Engineering Science, University of Essex, Telecommunications System Group, 1978.

108. P.S. Thiagarajan: *Elementary Net Systems.* In: W. Brauer, W. Reisig, G. Rozenberg (eds.): Petri Nets: Central Models and Their Properties, Advances in Petri Nets 1986 Part I, Lecture Notes in Computer Science Vol. 254, Springer-Verlag 1987, 26–59.

109. M. Tofte: *Four Lectures on Standard ML.* Technical Report ECS-LFCS-89–73, University of Edinburgh, Department of Computer Science, 1989.

110. J. Toksvig: *DesignBeta. Beta Code Segments in CP-nets.* Computer Science Department, Aarhus University, 1991.

111. N. Treves: *A Comparative Study of Different Techniques for Semi-flows Computation in Place/Transition Nets.* In: G. Rozenberg (ed.): Advances in Petri Nets 1989, Lecture Notes in Computer Science Vol. 424, Springer-Verlag 1990, 433–452.

112. R. Valk: *Nets in Computer Organization.* In: W. Brauer, W. Reisig, G. Rozenberg (eds.): Petri Nets: Applications and Relationships to Other Models of Concurrency, Advances in Petri Nets 1986 Part II, Lecture Notes in Computer Science Vol. 255, Springer-Verlag 1987, 218–233.

113. A. Valmari: *Stubborn Sets for Reduced State Space Generation.* In: G. Rozenberg (ed.): Advances in Petri Nets 1990, Lecture Notes in Computer Science Vol. 483, Springer-Verlag 1991, 491–515.

114. A. Valmari: *Compositional State Space Generation.* Proceedings of the 11th International Conference on Application and Theory of Petri Nets, Paris 1990, 43–62.

115. A. Valmari: *Stubborn Sets of Coloured Petri Nets.* Proceedings of the 12th International Conference on Application and Theory of Petri Nets, Aarhus 1991, 102–121.

116. J. Vautherin: *Parallel Systems Specifications with Coloured Petri Nets and Algebraic Specifications.* In: G. Rozenberg (ed.): Advances in Petri Nets 1987, Lecture Notes in Computer Science, Vol. 266, Springer-Verlag 1987, 293–308.

117. K. Voss: *Nets in Data Bases.* In: W. Brauer, W. Reisig, G. Rozenberg (eds.): Petri Nets: Applications and Relationships to Other Models of Concurrency, Advances in Petri Nets 1986 Part II, Lecture Notes in Computer Science Vol. 255, Springer-Verlag 1987, 97–134.

118. K. Voss: *Nets in Office Automation.* In: W. Brauer, W. Reisig, G. Rozenberg (eds.): Petri Nets: Applications and Relationships to Other Models of Concurrency, Advances in Petri Nets 1986 Part II, Lecture Notes in Computer Science Vol. 255, Springer-Verlag 1987, 234–257.

119. Å. Wikström: *Functional Programming Using Standard ML.* International Series in Computer Science, Prentice-Hall, 1987.

Index

Monographs in Theoretical Computer Science – An EATCS Series

C. Calude
Information and Randomness
An Algorithmic Perspective

K. Jensen
Coloured Petri Nets
Basic Concepts, Analysis Methods and Practical Use, Vol. 1
2nd ed.

K. Jensen
Coloured Petri Nets
Basic Concepts, Analysis Methods and Practical Use, Vol. 2

K. Jensen
Coloured Petri Nets
Basic Concepts, Analysis Methods and Practical Use, Vol. 3

A. Nait Abdallah
The Logic of Partial Information

Texts in Theoretical Computer Science – An EATCS Series

J. L. Balcázar, J. Díaz, J. Gabarró
Structural Complexity I
2nd ed. (see also overleaf, Vol. 22)

M. Garzon
Models of Massive Parallelism
Analysis of Cellular Automata and Neural Networks

J. Hromkovič
Communication Complexity and Parallel Computing

A. Leitsch
The Resolution Calculus

A. Salomaa
Public-Key Cryptography
2nd ed.

K. Sikkel
Parsing Schemata
A Framework for Specification and Analysis of Parsing Algorithms

Former volumes appeared as EATCS Monographs on Theoretical Computer Science

Vol. 5: W. Kuich, A. Salomaa
Semirings, Automata, Languages

Vol. 6: H. Ehrig, B. Mahr
Fundamentals of Algebraic Specification 1
Equations and Initial Semantics

Vol. 7: F. Gécseg
Products of Automata

Vol. 8: F. Kröger
Temporal Logic of Programs

Vol. 9: K. Weihrauch
Computability

Vol. 10: H. Edelsbrunner
Algorithms in Combinatorial Geometry

Vol. 12: J. Berstel, C. Reutenauer
Rational Series and Their Languages

Vol. 13: E. Best, C. Fernández C.
Nonsequential Processes
A Petri Net View

Vol. 14: M. Jantzen
Confluent String Rewriting

Vol. 15: S. Sippu, E. Soisalon-Soininen
Parsing Theory
Volume I: Languages and Parsing

Vol. 16: P. Padawitz
Computing in Horn Clause Theories

Vol. 17: J. Paredaens, P. DeBra, M. Gyssens, D. Van Gucht
The Structure of the Relational Database Model

Vol. 18: J. Dassow, G. Páun
Regulated Rewriting in Formal Language Theory

Vol. 19: M. Tofte
Compiler Generators
What they can do, what they might do, and what they will probably never do

Vol. 20: S. Sippu, E. Soisalon-Soininen
Parsing Theory
Volume II: LR(k) and LL(k) Parsing

Vol. 21: H. Ehrig, B. Mahr
Fundamentals of Algebraic Specification 2
Module Specifications and Constraints

Vol. 22: J. L. Balcázar, J. Díaz, J. Gabarró
Structural Complexity II

Vol. 24: T. Gergely, L. Úry
First-Order Programming Theories

R. Janicki, P. E. Lauer
Specification and Analysis of Concurrent Systems
The COSY Approach

O. Watanabe (Ed.)
Kolmogorov Complexity and Computational Complexity

G. Schmidt, Th. Ströhlein
Relations and Graphs
Discrete Mathematics for Computer Scientists

S. L. Bloom, Z. Ésik
Iteration Theories
The Equational Logic of Iterative Processes